THE PSYCHOLOG
TEMPORA

GW01452666

Understanding human temporality is a theoretical and experimental challenge, requiring oblique and imaginative ways to capture the manner in which time enters and structures human experience. This book bridges music, physics, and experimental psychology to present a unique perspective on the experience of time that is rooted in both physical theory and Gestalt psychology. Featuring a novel framework based on the idea that time enters mind through the dynamics of memory decay, it draws on the author's research experience in astrophysics and cognitive psychology to present a unique perspective on this topic. It will be of interest to students, researchers, and anyone seeking deeper insight into how the mind and body interact to shape personal experience and the world we inhabit.

DAVID GILDEN is a professor in the Department of Psychology at the University of Texas at Austin, USA. Trained originally in theoretical astrophysics, his research interests cover topics in mathematical psychology, attention, perception, and memory. Gilden is best known for the discovery of $1/f$ noise in human cognition.

THE PSYCHOLOGY
OF HUMAN TEMPORALITY

DAVID GILDEN

University of Texas, Austin

CAMBRIDGE
UNIVERSITY PRESS

CAMBRIDGE
UNIVERSITY PRESS

Shaftesbury Road, Cambridge CB2 8EA, United Kingdom

One Liberty Plaza, 20th Floor, New York, NY 10006, USA

477 Williamstown Road, Port Melbourne, VIC 3207, Australia

314–321, 3rd Floor, Plot 3, Splendor Forum, Jasola District Centre,
New Delhi – 110025, India

103 Penang Road, #05–06/07, Visioncrest Commercial, Singapore 238467

Cambridge University Press is part of Cambridge University Press & Assessment,
a department of the University of Cambridge.

We share the University's mission to contribute to society through the pursuit of
education, learning and research at the highest international levels of excellence.

www.cambridge.org
Information on this title: www.cambridge.org/9781009408783
DOI: 10.1017/9781009408776

First published 2026

Front cover: Uncharted Waters. India. From 'Theatrum Orbis Terrarum', by Abraham
Ortelius (1527–1598). Hand-coloured engraving. Published, Antwerp, 1592.
Christie's Images / Bridgeman Images

A catalogue record for this publication is available from the British Library

A Cataloging-in-Publication data record for this book is available from the Library of Congress

ISBN 978-1-009-40878-3 Hardback
ISBN 978-1-009-40876-9 Paperback

Contents

Preface

This book is about an approach to understanding human temporality through the prism of perceptual organization, what is known in the field of psychology as Gestalt. The book is, in consequence, a little odd as Gestalt is, by its very nature, indescribable except in the vaguest and most oblique terms. Nevertheless, this approach will require understanding what Gestalt is, and particularly how it operates in time. This understanding will come from three places: existential philosophy, experimental psychology, and the physics of dynamical systems.

Gestalt is about relatedness and what ensues from coming into relation. A good place to learn about relatedness is Kierkegaard's *The Sickness unto Death*. From the second paragraph:

> In the relation between two, the relation is the third term as a negative unity, and the two relate themselves to the relation, and in the relation to the relation; such a relation is that between soul and body, when man is regarded as soul. If on the contrary the relation relates itself to itself, the relation is then the positive third term, and this is the self. (Sören Kierkegaard, *The Sickness unto Death*, 1840)

Much of this book is about how temporality enters the life of mind when the relation is the positive third term and the self is realized. As might be clear from this bit of existential philosophy, the concept of relation is beyond the reach of the kind of discourse that explains things. Consequently, being in relation is demonstrably not a fit subject for a scholarly work, and as a further consequence, this book is not scholarly. That being said, this book will build out a theory of temporality from being in relation, from the positive third term.

One thing that distinguishes this book from *The Sickness unto Death* is that it does not proceed from pure reason, but is written from the perspective of an experimental psychologist. An experimental frame of mind was required because there were several points in the development of the theory where conjectures appeared. It is truly a serendipity that we are living in an era some 140 years removed from the birth of psychology. We inherit a collection of experimental methods and techniques that, on occasion, can decide the truth of a conjecture. Consequently, there is quite a lot of experimental psychology in this book. The emphasis on experiments and data is not of the scholarly sort where an entire literature is covered. No, here only a few relevant experiments will be discussed, and they will be discussed in detail. Most of the experiments, especially those in the later chapters, require no special knowledge or equipment, and could be conducted by anybody. So, whoever is interested in embarking on their own investigations is certainly welcome to see what they can add (or subtract).

In addition to existential philosophy and experimental psychology there is also some physics. The physics is important because it introduces ideas that do not seem to exist in psychology, and which are needed to understand how the mind builds bridges in time. The physics that is introduced is limited, having only to do with expanding our thinking about what memory is and how memory works. Regardless, the physics is described realistically and that means that there are a few equations. Equations are a particularly useful form of discourse because they speak plainly, and they make it possible to be specific and exact. Specificity and exactness will prove to be very useful in understanding how being in relation works in time.

Acknowledgments

The author wishes to acknowledge some of the people who in one way or another contributed to this work. Gary Aamodt was a mentor who taught me (and other people) the art of both/and thinking. Dennis Proffitt was a mentor who taught me (and quite a few other people) how to do psychology. Tanya Feinstein convinced me that I should give serious thought to writing this book. That had never occurred to me. Laura Marusich and Llewyn Paine were my first collaborators in the study of time. The empirical work on phase transitions in Gestalt owes much to Laura's dissertation on the contraction of time in attention deficit hyper-activity disorder. Llewyn's work on the segmentation of time-based groups was enormously instructive in developing the Gestalt perspective of this book. Tom Thornton was my collaborator in the study of $1/f$ noise, a dynamical form of memory. Much of the theory in this book developed from our work together. Taylor Mezaraups has been my collaborator for the past decade. First as an undergraduate, and then as a graduate student. Together we wrote the articles that form the empirical backbone of this book. Katherine Gonzalez Cuetara has been my most recent collaborator. She conducted the final speech study, showing amazing stamina in being able to listen to celebrities talk about themselves. Finally, I would like to acknowledge my daughter, Josephine Iacarella, who for many years has helped me keep things in perspective.

CHAPTER 1

The Nature of Sensation and the Experience of Time Passage

The branch of psychology that is concerned with the nature of thought, cognitive psychology, suffers from at least two fundamental problems. One problem is that nobody knows what a mind is. The second is that, even with the certain knowledge that minds exist, there is no principled theoretical framework that informs on how they might be studied. And then there is the oddity that the idea that the mind might be systemically studied is a very recent invention. Although people, philosophers and poets mostly, have been describing mind from the point of view of what they intuit, the idea that data might be informative apparently did not occur to anyone until the late nineteenth century. This situation creates a certain informality where the few people who are interested in the systematic study of mind are basically left to their own devices. When people are left to their own devices, the science that evolves will be guided in the first place by what people notice and then by what they deem to be important. The issue of what comes to be observed is especially relevant to this inquiry into human temporality as it begins with the proposition that the field of cognitive psychology has, for the most part, been valuing the wrong things.

The circumstances that flow from the informality of noticing and the vagaries of valuation have led to a perspective on temporality where a single phenomenon has come to be prized in importance. That phenomenon is that *humans and other animals can, with some accuracy, judge the duration of time passage.* This phenomenon might not have arisen as something to be noticed and valued were there not a pre-existent interest in processes of judgment. Psychology has its own peculiar history, and it happens to be the case that the earliest psychological investigations were often focused on processes of judgment, and particularly on judgments of sensory qualities such as brightness and loudness. Although temporal duration is not associated with a sensory quality (there are no receptors in the body for time), the experience of temporal duration came to be understood as belonging in the mix with brightness or loudness when it came to judgment. Nevertheless,

1

this book will proceed by noticing and valuing other things, things that do not involve judgement, which will make it clear that the experience of time passage is not at all like the experience of light and sound, and that presuming that it is leads to nonexplanatory theories of our time sense, and to incorrect interpretations of otherwise legitimately conducted experiments. Mostly it leads to an understanding of temporality that is in error.

These statements are not simple matters of fact, and they need to be justified. It is important to understand why an interval of time cannot be treated as offering a sensory quality and why the psychology that flows from that treatment is misleading. To properly motivate the Gestalt perspective on time taken here, it is necessary to first understand what it is replacing. There are specific structural assumptions that must be made about the experience of temporal duration for it to be treated as a sensory experience that might be judged, independent of whether those judgments are accurate or not. Our point of departure is an examination of these assumptions. Once we have clarified what it means to be an object of judgment, an illustrative example of a judgment experiment will be offered. This is a field that is driven by empirical inquiry and some specific knowledge about methodology and technique will be invaluable for appreciating what kinds of things are learned in judgment experiments. There are, in addition, theories and computational models that have been constructed to explain how it is that animals are able to judge durations in the first place. The most influential of these are pacemaker-accumulator models. These models deserve particular focus as they are quite transparent both about their psychophysical roots and the way in which time is conceptualized as a psychological quantity. Having waded through some representative methodology and some of the theory, the sensory perspective on temporal duration will be interrogated by returning to the core assumptions about time that launched this field. At this point the foundational theme of this book will be introduced; that the experience of time is unique by virtue of having a phase transition. Experimental paradigms and theories of duration judgment that fail to recognize that there is a phase transition are missing the single most important fact about the experience of time. It is this circumstance that opens the field of human temporality to an inquiry based in Gestalt psychology.

The Psychophysical Perspective on Time

Psychology, as a unified and coherent discipline, began within an intellectual movement that from about 1890 on had been developing rigorous techniques for the study of mind. This movement, known as sensory psychophysics, was principally concerned with the most primitive of psychological experiences, with the experience of sensation. One of the issues dealt with in this field was how energy impinging on a sense was experienced as sensation. A prototypical inquiry in this field might establish the relationship between perceived brightness and light intensity or between perceived loudness and sound intensity. Time passage might seem an unlikely candidate to be swept up by sensory psychophysics insofar as time does not impinge on the senses like, for example, light or sound. Nevertheless, temporal duration does have a critical property in common with sound and light intensity that made it susceptible to psychophysical investigation. This property is that light intensity, sound intensity, and the durations of time intervals all have magnitude. Even though temporal duration is not strictly an intensity, because it has magnitude it inherits the mathematical properties of magnitude. One key property that magnitudes have is that they can be graded in terms of size and so can be ranked and sorted. The sorting property is so fundamental to the theory of measurement that it defines a class: continua that can be ordered based on magnitude form the class of *prothetic continua*. So, although the body receives time passage in a way that is quite different from the way it receives light and sound, the things being received all belong to the same measurement class. It is with this single observation that the study of timing in humans and other animals was set in stone.

The observation that durations have magnitudes, that they can be long or short, is not particularly trenchant and the class of prothetic continua is hardly exclusive. Most things that arise in everyday experience are characterized by magnitude and can be meaningfully compared in terms of size. Physical intensities obviously have this property but so do mathematical continua such as riskiness and likelihood. In fact, the few things that do not have magnitude are worth pointing out – they belong to the class of *metathetic* continua. Direction does not have magnitude, and east is not more or less than north. Different tastes do not have magnitude in the sense that nutmeg is not more or less than basil. A given taste can be faint or strong and, in this sense, given tastes are on prothetic continua. The same applies to color. Blue is not more or less than red insofar as the perception of color is organized on a color wheel. Anything arrayed on a circle cannot be greater or less than other things on the circle. But colors can be more or less saturated, so in terms of saturation a given hue generates a prothetic continuum.

Sensory psychophysics began with the recognition that subjective experiences also had magnitudes. Although this may be obvious, it is after all the basis of the pain scale and all Likert scales (typically 4- to 7-point scales on which just about everything is rated), it is nevertheless quite amazing. That subjective experience could have magnitudes that are coordinate with objective magnitudes is one of the more interesting outcomes of adaptive evolution. That we would have an abstract sense of magnitude is not something to be taken for granted and should be appreciated for its weirdness. In the objective world, the world of physical matter, the magnitudes that form prothetic continua have dimensions such as inches, seconds, liters, and so on. But in the mind, there are no dimensioned quantities, there is just thought. Nevertheless, people can assign magnitudes to their private and subjective experiences and can meaningfully compare them. Consequently, there are two places under heaven where there are prothetic continua: in the physical world of matter where continua are dimensioned, and in the mind where the experience of physical continua is dimensionless. It is here that sensory psychophysics becomes a science. If there is a prothetic continuum of magnitudes in the world that creates a prothetic continuum of magnitudes in the mind, then there might be lawful relationships between these two continua. The originality and profundity of this idea cannot be overstated. Humans had for millennia the self-reflection to ponder subjective experience as well as the mathematics to collect such reflections into laws. But the notion that there might be laws of thought is a very recent development, dating back no further than the birth of modern psychology around 1900. A law of thought with the world on one side, the mind on the other, and an equal sign between them is a heady thing to contemplate, but this is exactly what sensory psychophysics was built for.

Stevens' (1957) enduring contribution to sensory psychophysics is an eponymic law that relates stimulus intensity to perceived magnitude. This law is powerful in two senses. First it is powerful in that it was intended to be applied to the gamut of human experience, and its scope is, in fact, breathtaking. The second sense is that this law is literally a power law, a type of mathematical function that makes intermittent appearances throughout the history of psychology and which will reappear in this book in discussions of both memory and animal body size. Stevens' power law is a piece of mathematics that specifies how subjective experiential states are produced by objective states of the world. It has the form:

$$\psi(I) = k\,I^{\alpha}.$$

In this equation ψ is the perceived magnitude, a purely mental quantity that must be discovered through experiments in magnitude estimation, and I is a stimulus intensity that is obtained through a physical measurement; k is a constant that plays the important role of making the equation dimensionally correct – both sides must be dimensionless as mental states such as $\psi(I)$ do not have dimensions. The exponent α is the key variable in this equation in that different sources of stimulation will have different exponents. The exponent informs on whether the experience of a source is magnified ($\alpha > 1$), minimized ($\alpha < 1$), or veridical ($\alpha = 1$). It should be pointed out here, perhaps unnecessarily, that Stevens' law is an enormously impressive achievement. It essentially collects the totality of world experience, at least that part that can be attributed to a source intensity, into a single law with a single parameter.

Although Stevens' law certainly looks like a law of nature, it is not a law of nature in any usual sense. First there is the issue of how the intensity, I, is interpreted. To appreciate how I operates in Stevens' law it will help to look at a physical law. Consider Newton's 2nd law of motion, $F = ma$, force equals mass times acceleration. The three terms in Newton's law have a set meaning; what they are does not depend upon context. Forces, masses, and accelerations refer to the same quantities regardless of what is accelerating, what type of force is acting, and what has mass. Acceleration, for example, the a in $F = ma$, refers to a particular quantity and it will always have the dimensions of length per time squared. This is not true of the intensity in Stevens' law. Intensity is not a particular quantity, and its dimensions will vary depending on the physical source. The intensity of light is measured in lumens or lux and is a different kind of thing than the intensity of sound, which is measured in terms of watts per square meter. So, although the I in Stevens' equation looks like it refers to a property of things called *intensity*, it requires quite of bit of interpretation to understand how it functions.

Even more unusual, from a mathematical point of view, is the $\psi(I)$ on the left-hand side of Stevens' law. It is intended to be read as "the magnitude of the subjective experience of intensity," but subjective experiences are not just one thing. The subjective experience of light magnitude is different from the subjective experience of sound magnitude; brightness is plainly different from loudness. Yet despite the manifest differences that exist between the different dimensions of experience, there is an abstract sense of magnitude that cuts across them all. This abstract sense of magnitude is friendly to quantitation. So even though a bright light could not be more different than a loud sound, somehow people are able to create something like a 7-point Likert scale for both and report that this brightness is, say, a 3 out of 7, and that loudness is, say, a 6 out of 7. It is this abstract sense of quantitative magnitude that is intended by the notation $\psi(I)$.

Figure 1.1 illustrates three examples of Stevens' law and shows how three different types of intensity are experienced. More importantly, it implicitly illustrates how temporal duration has been historically conceptualized. As the experience of duration lies on a prothetic continuum alongside the myriad forms of sensory experience, duration can share space with the dimensions of shock and brightness in a plot of perceived intensity. The only way in which duration, brightness, and shock are individuated is through α, the power law exponent. In this regard duration is quite unusual. Shock, for example, with $\alpha = 3.5$, is experienced in a rapidly accelerating way (the absolute numbers are not meaningful here, but ratios and percentages are). A doubling of shock intensity is experienced as an 11-fold increase in the feeling of shock. In contrast, brightness with $\alpha = 0.33$ is experienced in a decelerating way. A 100% increase in light intensity is experienced as only a 25% increase in brightness. The duration curve is unusual because it is roughly linear. The duration exponent $\alpha = 1.1$, implying that a doubling of a duration in the world is experienced as a doubling of the subjective experience of duration. This might seem to imply that people experience the passage of time veridically.

Figure 1.1 tells the story of how the experience of time passage found a home within the discipline of sensory psychophysics. This homecoming had far-reaching consequences. Psychophysics is a powerful tool and here we should acknowledge what powerful tools can do. Once the passage of

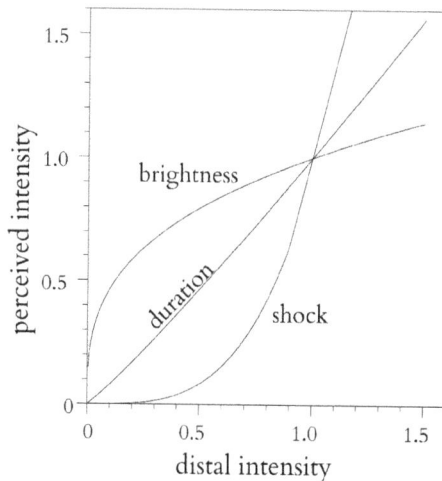

Figure 1.1

time was put into the mix with light and sound (and shock and many other sensory experiences), the die was cast for the unfolding of timing theory and for the conduct of experiments in human and animal timing.

The Peak Interval Procedure and Weber's Law

At this juncture it might be instructive to consider a concrete and productive psychophysical paradigm for the study of timing in animals. There are many paradigms that might be considered for inclusion, but the peak interval procedure is relatively straightforward and illustrates well how animals express their understandings of time passage. As the procedure involves a bit of unsubtle trickery, it would not work as well with human participants as it does with rats and pigeons. Nevertheless, the nature of the data produced is universal and would capture that nature of duration judgment in any animal that was capable of learning to respond to a given interval of time.

In this procedure animals are removed from their cages and placed into a conditioning chamber. At some point an environmental event is introduced that acts to signal the beginning of the time interval. Appreciating the environmental meaning of the signal and that a time interval is involved is something the animal learns over many episodes of being put in the conditioning chamber. The quality and type of the signal is subject to the whim of the experimenter and consists typically of the onset of a sound or light that clearly marks the beginning of an epoch. It is not a trivial observation that animals, people included, can appreciate that lights and sounds might signify – have meaning beyond their physical characteristics. In this paradigm, duration is measured as elapsed time following the orienting signal.

Duration is introduced as something that is environmentally relevant by training the animals on a fixed interval (FI) schedule. In FI training animals learn that after FI seconds have passed, a response will be followed by a food or juice reward. Typical FI values are in the tens of seconds, presumably because such times permit good resolution of response rate – the dependent variable (the thing measured) in these studies. Different animals respond differently, rats press bars and pigeons peck, but both bar presses and pecks are discrete events. As discrete events, they can be counted and so lend themselves to rate measurement. Various types of laboratory animals, it turns out, can learn the relation between reward expectancy and FI duration, and they communicate this through their response rates (bar presses per minute or pecks per minute). As the FI

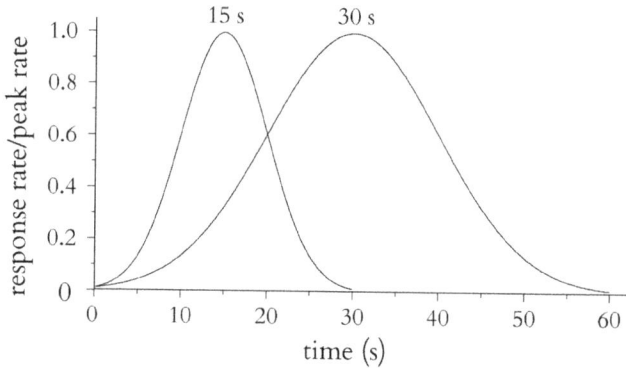

Figure 1.2

duration approaches there is an increasing amount of bar pressing or pecking. The trickery is that on some trials, referred to as *probe* trials, no reinforcement is given. The animal is basically stood up, but they do not stop responding after the FI has passed. In this paradigm the shape of the response curve tells the story of what animals understand about quantities of time passage.

Figure 1.2 illustrates idealized rate curves on probe trials for an animal that has been trained on two fixed intervals, 15 s and 30 s. In practice the way data such as these would arise is that the onset of a light, say, would signal that a reward would be available after an FI of 15 s, whereas the onset of a sound (white noise burst typically) would signal that a reward would be available after an FI of 30 s. The x-axis then is the amount of time that has passed since the presentation of one of these signals. The y-axis is the rate at which the animal is responding using whatever body part is relevant – rats push bars and pigeons peck plates. This animal demonstrates its ability to learn the fixed intervals by generating a peak response rate at 15 s when a light turns on and a peak response rate at 30 s when a sound commences.

The response rates in both conditions follow what appears to be bell-shaped curves. To be sure, bell-shaped curves are not always observed on probe trials, but they are typical. It may be necessary to clarify that the appearance of a bell-shaped curve in response rate has nothing to do with the normal (or Gaussian) distribution that is a fixture in the analysis of data. The response rate curve is not a distribution, it is just a graph of response rate. To the extent that the response curve is bell-shaped, it illustrates that the growing excitement engendered by an approaching FI duration is mirrored by a growing disappointment that the time for reward delivery has passed.

The response rate curves tell a more nuanced story than just that this animal can learn fixed time intervals. The width of the response rate curve, however it is defined, provides a measure of uncertainty. That there is uncertainty is evident from the fact that there is quite a bit of bar pressing or pecking both before and after the FI. The range of times when the response rate is relatively high defines an acceptance zone, essentially a measure of FI-ish, close to but not quite the FI. What it means for a response rate to be "relatively high" is typically defined by rates that are higher than half of the peak rate. The acceptance zone would then be the full width at half maximum (FWHM). The FWHM provides a quantitative estimate of timing uncertainty in each FI condition.

Where the peak interval procedure makes deep contact with the psychophysics of judgment is in the way the magnitude of timing uncertainty scales with the magnitude of the FI. In these idealized data the FWHM at an FI of 30 s is about twice that of the FWHM at an FI of 15 s. That is, 30 s-ish is about twice that of 15 s-ish. This kind of scaling is our first encounter with Weber's Law. There are several ways of expressing Weber's law and here the appropriate formulation is:

the uncertainty of experience ~ the magnitude of experience.

There are innumerable instances of this general law and several different ways in which uncertainty is expressed and experienced. In the context of duration estimation, it leads to the width of acceptance zone, the width of what is "ish," being proportional to the size of the time interval that has been trained.

The proportionality in Weber's law is both a profound and completely common aspect of human experience. It is profound in that it speaks to the fundamental nature of experienced magnitude. Even though it is not a physical law, it has the character of a physical law – almost like an uncertainty principle. It is common in that the proportionality is found in all sensory systems and in magnitude judgment generally. To get a sense of its generality, consider how an inch might be demonstrated with the thumb and index finger. No one's inch will be an inch – it will be a little too big or a little too small. This is the acceptance zone, the inherent uncertainty, of an inch. When the hands are placed apart to indicate the size of 12 inches or a foot there will again be error. But the error is now much larger. When we indicate how big an inch is, we will not be off by an inch, but we can easily be off by an inch when we show how big a foot is. The error always grows with the size of the thing that is judged, estimated, or felt. If the error grows proportionally with the size of the thing judged, then the system of judgment is said to satisfy Weber's Law or be Weberian.

That animals can accurately judge time passage and respond appropriately to an FI is certainly evidence that animals can encode, store, and retrieve the durations of time intervals. That the process of time judgment has the Weberian property is evidence that time judgment is not materially different than the judgment of sensations generally. Both conclusions will be challenged shortly where it will become clear that although the data in the peak interval procedure may be unimpeachable, the interpretation given to the data is not. There are good reasons to suspect that it is not time itself that animals are judging when they reach maximum pecking and/or pressing after 15 or 30 s has elapsed.

Pacemaker Theory of Timing

When an animal starts pressing a bar or pecking on a plate as a fixed interval of time approaches, it is doing something quite remarkable. Duration does not have a sensory quality. There is no energy associated with duration and there is no sensory organ that allows animals to perceive time. Yet the animal is aware of temporal displacement, that time is passing. Logically, the animal must have some form of memory that allows different moments of time to be differentiated. The question that arises then is what kind of memory allows humans and other animals to mark specific intervals of time passage. Scalar expectancy theory (SET) is one highly influential attempt at constructing a memory system that can measure the passage of time (Gibbon, 1977; Gibbon, Church, & Meck, 1984).

In the most elementary terms, SET is a scheme for making $\psi(I)$ out of I for the kind of intensity that is temporal duration. It achieves this through the artifice of constructing a brain-based clocking mechanism (Gibbon, Church, & Meck, 1984). The logic of the SET clock is illustrated in Figure 1.3.

SET begins by presuming that there is a source of continuous oscillation in the brain that may be harnessed to produce a sense of time passage. Oscillations in physics are generally sine waves and that is what is depicted here. The details are in the harnessing mechanisms insofar as brain oscillations, brain waves, do not communicate the duration of time intervals. To transform oscillations into something that has magnitude, SET makes a second presumption – that there is a mechanism that can pick out phases of the oscillation to create a train of pulses. Any point in the repeating cycle

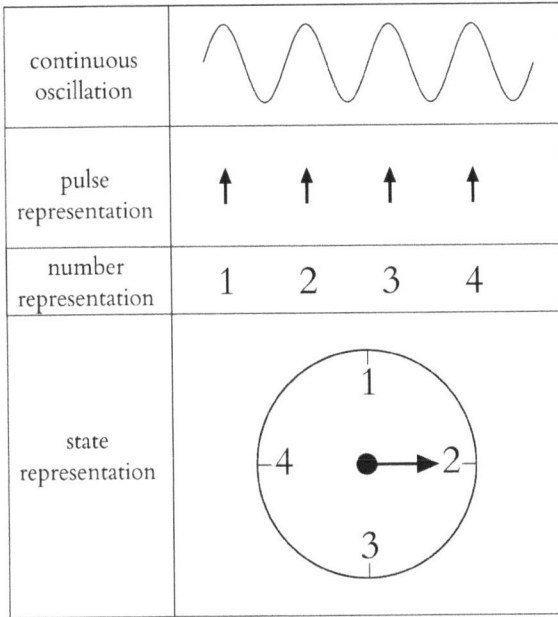

continuous oscillation	
pulse representation	
number representation	
state representation	

Figure 1.3

will serve this purpose, and in the second panel of Figure 1.3 crests (wave maxima) are used to create a pulse train. The pulse train is illustrated as a set of separated arrows to highlight that a pulse is a discrete thing. What a pulse train has that a continuous wave does not is numerosity; because pulses are discrete, they may be counted. So, in the third panel of Figure 1.3 the pulses are counted, and the four arrows are now replaced by the numerals 1 through 4. It is unclear how pulses might be counted in a biological system, but presumably it is accomplished through a state representation where pulse arrivals are attended by transitions of state. The fourth panel is an idealization of how states embody number. In this idealization there is a pointer that moves one step clockwise with the reception of each pulse. The position of the pointer is the embodiment of duration magnitude. Regardless of whether any of this has biological plausibility, this picture has for many years been influential in theories of interval timing.

SET supplements the scheme for turning oscillations into magnitudes with a switch and a set of buffers that allow humans and other animals to behave in ways that are meaningfully related to time passage. The full

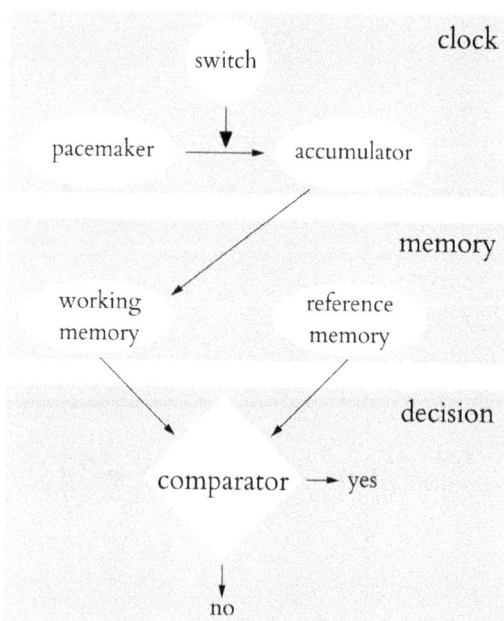

clock

switch

pacemaker ——▼——▶ accumulator

memory

working   reference
memory   memory

decision

comparator ——▶ yes

no

Figure 1.4

theory is illustrated in Figure 1.4 as a kind of flow chart that is intended to indicate the process in which counts are prepared, accumulated, and then compared to environmentally relevant counts that have been learned in the past. The switch is of special interest because it is the locus where the internal clock meets the world. Logically, for this scheme to operate as a counter that can accomplish duration estimation, there must be a zero count when the interval of interest to the animal begins. Beginnings and zero counts are negotiated in SET through switches. The switch is assumed to have two states. In the open state the oscillator is running but it does not pass pulses to the accumulator. When something in the environment tells the human or other animal that an interval of interest is beginning, the switch moves to the closed state and pulses are accumulated.

The remainder of SET is concerned with how meaningful behavior might be produced by counting pulses. Meaningful behavior is conceptualized as being produced when the count arrives at some target. SET requires the existence of target counts to operate as a timing mechanism and so there is the supposition of a buffer, referred to as the reference memory, that holds special counts that the animal has learned through prior experience. For example, if a rat is being assessed in a peak interval procedure, it may have learned that after 20 s has passed a bar press will lead to a food reward. The rat does not know anything about seconds, but 20 s corresponds to some count, a count that presumably reflects the frequency of the oscillator and the rate at which pulses are produced. As an example, a 40 Hertz (Hz) oscillator produces 40 pulses per second and 800 pulses in 20 s. In this example, reference memory would contain a state corresponding to 800 pulses.

The final steps of SET describe a scheme for how the progressing count might be acted upon. Another buffer will be required to store the ever-increasing counts that correspond to the ever-increasing passage of world time. The accumulator might be conceptualized as a working memory (it is in some depictions of SET), but here it is regarded as just being a counter. It passes count states to a working memory that is not regarded as a counter but as a repository for whatever the accumulator passes to it. Finally, the state of the reference memory and the state of working memory are continuously compared, and when there is some degree of equality between the reference and working memory states, the relevant duration is deemed to have passed, and an action is invited. The invitation is denoted by the "yes" in the figure.

Digital and Analog Clocks

Scalar expectancy theory is an explicit computational theory of the time-keeping sense in humans and other animals. To this end it succeeds in providing an account of how, say, a rat, person, or pigeon could learn to track and respond to the passage of an interval of time. Its principal virtues, that it is explicit and computational, are also potential faults. It is also, oddly, a close representation of how quartz clocks keep time. Perhaps it is merely a coincidence that SET and its associated timing model (Figure 1.4) followed soon after Casio introduced the first mass-produced quartz wrist-watch. Placing the inner workings of a quartz clock into the head does succeed in creating a head that can keep time, but it also suggests that timing theory might benefit from a larger perspective.

That the pacemaker-accumulator component of SET chops a continuous analog signal into pulses for the purpose of counting might create the impression that continuous analog signals are not appropriate for timekeeping. This is not true; counting is not required for timekeeping. Putting aside the difficult question that SET was designed to solve, how humans and other animals can accurately judge the durations of time intervals, it is important and necessary to point out that there are many expressions of timekeeping that are analog in nature. Here a single example will make the point that analog timing mechanisms proliferate through animal bodies and through nature generally.

The example that will be considered here is the decay process. Arguably, decay is the most generic process in nature as every perturbation away from a state of equilibrium will involve a decay process as the equilibrium state is restored. Normally decay is not considered to be a mechanism of time-keeping, but that is only because physical processes are typically conceptualized as functions of time. In this conception the evolution of time produces system evolution. But turning this perspective on its side will create a clock. How this works is illustrated in Figure 1.5. On the left a decay process is illustrated in the way that decay is usually pictured. Time is on the x-axis and system states are on the y-axis. This plot gives the strong visual impression that as time flows, system states change. In this orientation, the interpretation is that system states are inferred from time. Given a measurement of time, t, the system state, x, is read out.

Now the plot will be transformed to a less familiar orientation, shown on the right-hand side of Figure 1.5. In this orientation the system states are on the x-axis and time is on the y-axis, giving the impression that time is inferred from system states. In this way of conceptualizing the decay process, we are not given time, but instead we are given a state of the system, and the state specifies the time. Although the physical process of

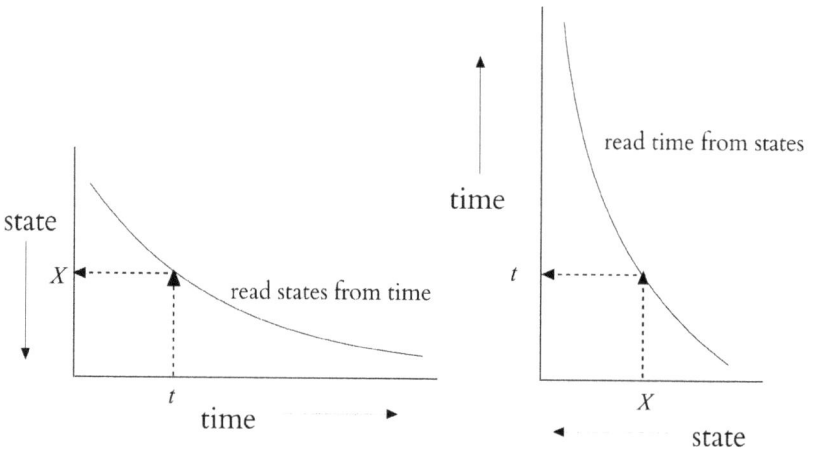

Figure 1.5

decay will be less familiar than a clock face, they do operate in the same way. Reading a clock involves looking at the hand states and inferring the time. And to be clear, everybody is also quite familiar with inferring time from physical states: If you are hungry your body may be telling you that it has been a while since it was fed. If you are tired, consider that your body has been clocking your time awake and building sleep pressure.

The general situation with decay processes is that they can be construed as clocks only up to a point where the system closes in on its final resting state. There will always be a point at which time moves forward, but the decaying system does not make noticeable progress in producing new lower states. Eventually the process asymptotes to what mathematicians call an accumulation point. As decay proceeds, states become increasingly close to each other and therefore become increasingly less useful for the purpose of reading time. This behavior can be visualized in the right-hand side of Figure 1.5 where the curve becomes essentially vertical – change in time without change in state. Because decay processes inevitably asymptote, a decay clock does not have the range of SET. Scalar expectancy theory can provide a pulse count for any interval of time so long as the oscillator churns out pulses. A decay process is clocklike only for the early period when states are not piling on each other. The infinite range of SET might be viewed as a virtue, a good thing, if we think that the scope of animal timekeeping is unbounded. If, on the other hand, animal time-keeping turns out to be bounded, then SET is supplying animals with a clock that they cannot use.

The Deep Structure of the Duration Continuum

The idea that the durations of events are plausible things for humans and other animals to judge is supported by the fact that humans and other animals manifestly do succeed in doing just that. It follows that it is meaningful to inquire as to how judgment is accomplished (say, through pacemaker-accumulator models), and it is meaningful to examine the psychophysical properties of the judgment process (like in the peak interval procedure). This fact structure makes a strong case that asking humans and other animals to judge event durations leads to productive and transparent science. The truth is, however, that this issue is far from settled, and it only appears to be so because the perspective that judgment offers is apparently not capable of recognizing that humans and other animals are uniquely ill-suited for judging the duration of events. A slight change in perspective will create an entirely different story. In that story the perception of time fails to form the kind of experiences that would allow durations beyond a couple of seconds to be apprehended as objects of judgment. This will, of course, necessitate a reevaluation of what was hitherto taken to be as obvious and manifest.

To understand what it means to experience a temporal duration it will be helpful to go back to Stevens's law and examine what the law assumes about the nature of sensory experience. Recall that Stevens' law is a relation between two things – between intensities, I, in the world and the subjective experience, $\psi(I)$, of those intensities. More fundamentally, Stevens' law articulates a kind of commonsense assumption about how mind is situated in the world. If a tree falls in the woods and there is an animal there also, does it make a sound? Of course it does. For every sound produced in the world in the presence of an animal with ears, there is an experience of that sound. The same comment applies to all sensory experience. The world impinges on the senses and experiences of the world ensue. That experiences attend sensory stimulation is practically what is meant by the "equal" sign that connects I to $\psi(I)$ in Stevens' law. This is what it means to have sensory experience, that for every I there is a $\psi(I)$. It is the handshake between intensities and the experience of intensities that allows Stevens' law to be written and for it to make sense. All of this is obvious, and that is the problem because there is one application of Stevens' law where this handshake fails.

Let's now apply Stevens' law to the durations of events with a bit more care than was given in constructing Figure 1.1. It is important to be as clear as possible here. The "equal" sign in Stevens' law entails that for every interval of world time there is a corresponding experience of the duration of that time interval. This statement seems so straightforward that it seems impossible that it could not be true. If that statement is acceptable, then so is the Stevens' law plot of experienced duration shown on the left-hand side of Figure 1.6 (replotted from Figure 1.1). The plot does nothing more than give shape to the idea that the passage of time in the world is experienced. This is what allows a line to be drawn that connects world time to experienced time.

The point of departure of this book is that the left-hand side of Figure 1.6 does not paint a true picture of the experience of time passage. The true picture is depicted on the right-hand side. On the right-hand side, the Stevens' law plot is interrupted by a dragon, marking the end of the known world. Dragons are appropriate here because it is not true that for every world duration there is an experience of that duration. Although it is true that there is no end to what duration an event might occupy, there are

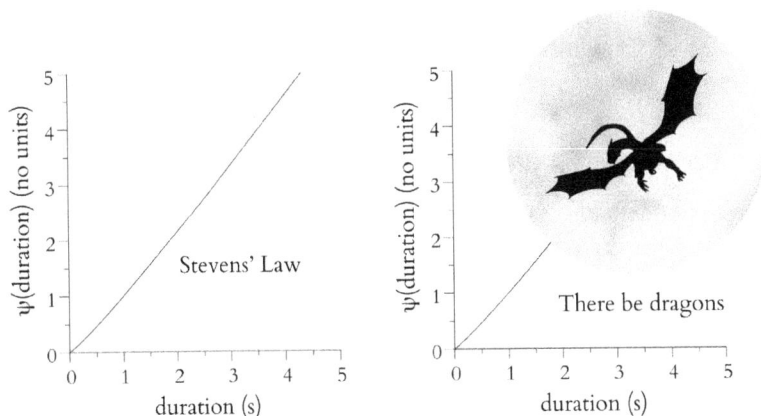

Figure 1.6

definite limits to what durations may be experienced. The *y*-axis, the axis of $\psi(I)$ in fact ends after a few seconds. *Beyond a few seconds there is a phase transition where the experience of duration essentially ends; there is no $\psi(I)$ for durations, I, greater than a few seconds.* In early cartography dragons were placed at the edge of the known world, and here dragons replace Stevens' law at a couple of seconds. This is what a phase transition in time means: In the first few seconds of time passage there is a rich experience of time, so rich that it is felt. But beyond a few seconds, the passage of time becomes uncharted territory.

The perspective that leads to dragons flying into Stevens' law is radically different from the judgment-driven perspective that regards time passage as just one more prothetic continuum, just one more dimension of experience. The perspective that sees dragons is known as Gestalt psychology. Gestalt is, to put it mildly and without prejudice, a very different kind of psychology than sensory psychophysics. Whereas sensory psychophysics is quite sophisticated and lives at the intersection of physics, biology, and mathematics, Gestalt, in contrast, is quite unsophisticated and lives at the intersection of philosophy, dialectic, and ecology. Nevertheless, Gestalt provides a conceptual framework for understanding the nature of human experience, and it provides the framework for understanding what is really going on with the experience of time passage.

The road to seeing dragons will not be long. In the next chapter the basic ideas that flow from Gestalt psychology will be introduced. Then in the chapter following, the phase transition in temporal experience will be introduced. The rest of the book will be about why there is a phase transition in human (and presumably mammalian) temporality and what the implications are for human temporality.

Perspective on the Psychophysics of Duration Judgment

There is undeniable evidence that humans and other animals can accurately judge the durations of time intervals. This evidence does not seem to be consistent with the Gestalt perspective on time, which insists that beyond durations of a few seconds there is no coherent or sensible experience of time. Evidently there is a distinction to be made here. It may not seem to be an important distinction, but it is central to understanding temporality. The distinction is between time passage, the t in the physical description of the world, and the history of events that occurs in consequence of time passage. Time and history are not the same thing, and they must be distinguished.

In psychology experiments that have to do with the judgment of time intervals there is the belief that it is time, physical time, that is being judged. This belief is explicitly reflected in the structure of SET. The core feature of SET is a pacemaker that is responding to the passage of physical time. The counts that are sent to the accumulator are a record of the passage of physical time and the counts that are stored in reference memory are records of previous passages of physical time. Literally every component of SET is based on physical time and just physical time. Put most simply, SET operates in terms of the time, t, that appears in physical descriptions of the world. Now, there is no question that animals may behave as if they are judging durations based on physical time, but if we deny that it is physical time that is being tracked, the need for SET and all similar theories of timing disappears. To be clear, duration judgment does require some form of memory. The proposition here is simply that a clock-based memory that stores physical time is not the right kind of memory.

History runs on a parallel track with time. On one track are moments, the stuff that makes time flow. On the other track are the things that happened at those moments. An animal that is aware of time passage may be tracking moments, perhaps by counting them in a clock manner as conceptualized by SET. Or the animal may just be aware of the flow of events that occurred before, say, a food pellet rolled out toward it. It happens that people and laboratory animals such as rats and pigeons do have memory for events, and this second track is always available. The two tracks are distinguished by their range: The existence of a phase transition in time does not affect the history track, it only affects the time track. So, although there may be no Stevens' law, no $\psi(I)$, for interval durations exceeding a few seconds, there is a continuous and virtually unlimited process of memory formation. Regardless of how time flows in experience and regardless of how that experience is structured by a phase transition, animals have a sense of their personal history, what is called their autobiographical or episodic memory. The proposition here is that episodic memory provides a continuum that can support duration judgment. Does not the narrative of our personal history keep us informed about interval durations – how long we have been looking for our keys, how long we have been waiting in line, how long it has been since we arrived at this stop light? In this view, experiments on duration judgment are not so much about the experience of time passage as they are about the experience of personal history. That leaves open the question of how time is, in fact, experienced. Answering that question is what this book is about. The point of departure is Gestalt.

CHAPTER 2

Gestalt and the Otherness of Everyday Experience

The theoretical position developed in this book is that time principally enters mental life through the portal of perceptual organization and particularly through the formation of groups. What perceptual organization is and how it works in space and time are not small topics. And they are not straightforward topics; it will take some work to truly understand that the world we perceive is a world that we create. It is very difficult to give up the notion that we are in close contact with reality. To get started, the focus will be on grouping in the spatial domain. The primary goal will be to gain some awareness of the active and creative role of mind in perceptual organization. This awareness will be essential in the following sections that take up grouping in the temporal domain.

Introduction to Gestalt

It is unfortunate that the first thing to be discussed, perceptual organization – or Gestalt as it has been historically referred to, is surely also the most elusive subject encountered in psychology. Gestalt is quite unlike any other topic in any science. It is a mostly informal discipline that has historically proceeded by the issuing of invitations; invitations to look or listen to something and then to think about what has been experienced. We will start off, then, by looking at a few discs, and attempting to be as aware as possible in thinking about what that experience consists of. This will also be a good introduction to looking and thinking in the absolutely simplest terms.

Figure 2.1 shows three panels, one with one disc, one with two discs, and one with three discs. This is a simple figure, but it contains virtually everything that is required to understand the scope and complexity of human perception. Let's start with the left panel containing one disc. There is a lot going on with that disc. First it is hard to see that disc, really see it, as being just a disc. That disc comes with a set of relations that

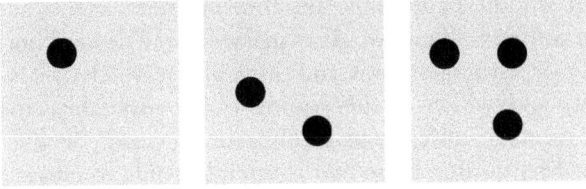

Figure 2.1

emanate from it and interact with relations created by the frame. The frame is unfortunate, but it was necessary to put the disc somewhere if it was to be included in the book. These relations create a place where the disc is located, in the upper left-hand corner of the frame. So instead of just having a disc, we now have a place and all the distance attributes like near and far that place entails. The frame is also interacting with the disc by creating shape contrast. The disc lives in a square world, a world that does not look like it. Also, the space it takes up seems to be structured by the square boundary. There are many places to put the disc, and the different places have different significances in the mind. Issues like this are central to photographic composition as the space within a frame is highly structured in terms of differential visual impact. Consequently, the most important part of any photograph is the frame. We are now at the beginning of Gestalt. In the physical world, things are located at positions, 3-tuples (x,y,z) in a coordinate system. In the mind of a perceiving animal, however, things occupy places. Occupying a place and being located at a position are not the same thing. Place is something that emerges from position in the minds of animals. Places have significance, including being sources of comfort, belonging, and status. Places are part of an environment that contains many places, forming a complex web of meaning – a world.

The middle panel contains two discs and now each disc is seen not only within the frame, but in relation to the other. Is it possible to see the discs in the middle panel as being by themselves, like the disc in the left panel? This is subtle. It seems that so long as you can see two discs, they provide company for each other and will never appear isolated. The discs in the middle panel form a group, a new type of thing that is not present in the left panel. Perceiving discs in relation to each other has a variety of consequences, what the Gestalt psychologists would call emergent properties – contents of the mind that are not in the physical stimulus. One key property is proximity. In the physical world there is distance, but in the mind, distance is transformed into near-by and far-away. The discs in the middle panel look like they are near each other. Another property is contour. These two discs make either a descending or ascending contour depending on how you think about which disc is the beginning point, and which is the ending point. Arguably, there are an indefinite number of emergent properties in any of the panels of Figure 2.1. Mentioning these two is intended only to create a mind-set where they may come into awareness.

In the right panel there are three discs, and now each disc has more company, there are more contours, and there are new emergent properties. A group of three things looks quite different from a group of two things. It is not that two and three are different numbers. That is not the issue. It is more that the quality of the group changes, and the way the discs occupy space has changed. Adding a third disc brings into being an entirely new entity, a triangle. Triangles are attended by any number of emergent properties, including the creation of interior regions and exterior regions, and the phenomenon of pointing. This triangle can point in three different directions, as can all triangles. It points up and to the left, up and to the right, and down and to the right – but not all at once. At any given moment it points just to one place.

There are two levels of description that are invited by Figure 2.1. One level is the physical description. This description is simple, clear, and straightforward. It consists entirely of a cataloguing of discs, where they are and how big. One of the most important properties of this description is that it is referential, the discs in the description can be pointed at. The second level is primarily what we are interested in here. This description is about qualities that are not referential in this sense but are still disc oriented. Gestalt attempts to describe the experience of discs being *together*, as forming a group. It is not possible to point at a group, and the existence of a group does not arise in a physical description. Try pointing at one of the groups in the middle or right panel. Where are you pointing? The togetherness that is a group comes into existence within the mind of a perceiving animal; minds create the togetherness of a group.

An experiential vehicle for understanding groups and grouping is the percept of multistability. Multistability invites a kind of mind-play where different grouping arrangements come into and out of existence, sometimes spontaneously and sometimes through acts of will. There are dozens of Gestalt demonstrations of multistability, some of them quite famous such as the Necker Cube and the duck/rabbit reversible figure. One of the most interesting things that attends the experience of multistability is the experience of fun. It can be (somewhat) amusing to experience the radical

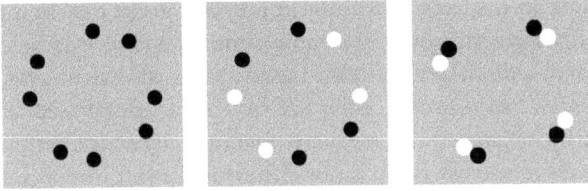

Figure 2.2

changes in perception as different grouping arrangements come into and out of awareness. Figure 2.2 provides a set of homemade examples that make the relevant points.

In the first panel eight discs have been drawn so that each disc has a close neighbor and a far neighbor. In both visual and auditory visual perception one of the strongest determinants of group formation is proximity; things that are close together will be perceived to go together. This is observed here as the discs that are closer to each other arrange into four separate pairs. The strength of proximity as a grouping force may be experienced by attempting to mentally rearrange the discs so that far neighbors are perceived to go together. This will be very difficult – perhaps impossible – as the mind will spontaneously pull the near neighbors back into pairings. The second panel is identical to the first except that now four of the discs are white and four are black. Discs that are closer to each other may still form pairs on the basis of proximity, but now the thing that was previously difficult is easy because far-neighbor discs that share a color also effectively form pairs. Color is just one of a myriad of features that may be used to promote grouping, and here it is chosen simply because it creates groups that literally pop into existence. In the third panel the grouping force of proximity is enhanced through contact, leading to highly salient groups of touching black/white pairs. Still salient are groupings by color, especially the percept of quartets – one group of four black discs and one group of four white discs. It is in this panel that the two opposing grouping forces are most vividly experienced.

Multistability arises from the circumstance that the pairs that form from proximity are not the same pairs that form from common color. There is within the mind a struggle between color and proximity as they actively compete for discs to include in their respective groups. A few moments spent experiencing the discs as they migrate from proximity pairs to color pairs will make clear that multistability is a little weird. One thing to notice is that grouping arrangements are experienced one at a time. There seems to be a prohibition on any given disc being collected into two different groups at the same time. If a disc is in a color-based group, then it is not in a proximity-based group, and vice versa, if a disc is in a proximity-based group, then it is not in a color-based group. This is noteworthy because there are many instances in which set membership is graded, where membership is not either/or. In fact, there is a branch of mathematics referred to as fuzzy set theory where elements may have degrees of membership. Fuzzy set theory captures the kind of thinking that is "ish", like in "youngish" or "oldish." One of the many things that multistability teaches is that perceptual grouping is not fuzzy, and that group membership is not graded. There does not seem to be "ish" in group membership. In the moment, a disc is never sort of in one group while also sort of being in another. Similarly, even though a triangle has the potentiality to point in three different directions, at any given moment it always points in one definite direction. The emergent property of direction is also not "ish".

A second thing to notice about this form of multistability is that it exists, that color similarity and proximity are somehow able to produce competing groups. Whatever mental property creates the sense of togetherness, it seems to act through the agency of proximity and color similarity in the same way. The situation here is reminiscent of the way forces work in physics. In Newton's 2nd law of motion, the F in $F = ma$ refers to any force regardless of what makes it. This circumstance is what allows forces to be added together and for systems to be brought into dynamical equilibrium. So, for example, gravity acts on the body to pull it toward the ground. Muscles can be used to exert forces to oppose gravity and accommodate sitting, standing, walking, and so on. Opposition is possible because force is force regardless of its provenance. In a similar sense, the gathering of things together into a group is not differentiated by the feature dimension that does the gathering. The gathering produced by common color is the same gathering that is produced by proximity. Because proximity and common color are doing the same thing, they can be brought into opposition. The opposition is experienced as multistability.

The Origins of Gestalt Psychology

Psychology is an odd science for many reasons, but maybe the oddest thing about it is that it did not emerge as a coherent discipline until the very end of the nineteenth century. Although it is the case that the nature of mind has been a topic of philosophical inquiry for millennia, subjecting psychological phenomena to systematic inquiry is a relatively recent enterprise. That one might develop a law of forgetting (Ebbinghaus), for example, or a law relating perceived sensation and stimulus energy (Wundt, Fechner), are very modern ideas. The European culture that created what is now recognized as the field of psychology also noticed that the formation of groups was occurring in perception, and in the early decades of the twentieth century there was a concerted effort to bring grouping within a systematic and fundamental theory. One of the principal contributors to this effort was Kurt Koffka. As the existence of this field, as a field, is largely due to him and to a small group of contemporaneous, mostly German scientists, it will be instructive to review his thoughts on what is going on with groups such as in Figures 2.1 and 2.2.

Koffka thought about the discs and groups in Figure 2.1 as existing at two different levels of structure; the discs exist as *parts* and the group exists as a *whole*. It is in this context that Koffka wrote something memorable, a few sentences that have resonated as a sort of meme for understanding what Gestalt is about:

> It has been said: The whole is more than the sum of the parts. It is more correct to say that whole is something other than the sum of its parts, because summing is a meaningless procedure, whereas the whole–part relationship is meaningful. (Koffka, 1935)

It may not be clear yet, but Koffka has given us exactly what we need to understand what happens when a group forms. Yet, one of the most interesting aspects of Koffka's comments about parts and wholes is that they were written in the first place. How does it come to pass that the contents of everyday experience need to be pointed out and explained by a Gestalt psychologist? Koffka's readers were not seeing for the very first time. They were not tourists of visual experience. The same thing could be said of this chapter: why would anybody presume to point things out about the qualities of everyday experience? It is one thing to point out facts about, say, rods and cones, as knowledge about them is arrived at through a true process of scientific investigation and discovery. It seems like it is quite another thing to point out that three discs may be seen as a triangle. This seems so obvious that it hardly requires mentioning. In this discussion, however, nothing will be taken as being obvious, and the more interesting question is how obviousness works.

Koffka is not the first person to wonder about wholes and parts, and there is a context for this distinction that goes back to Aristotle. In the *Metaphysics* (Book VIII, 1045a.8–10) Aristotle writes:

To return to the difficulty which has been stated with respect both to definitions and to numbers, what is the cause of their unity? In the case of all things which have several parts and in which the totality is not, as it were, a mere heap, but the whole is something beside the parts, there is a cause; for even in bodies contact is the cause of unity in some cases, and in others viscosity or some other such quality. And a definition is a set of words which is one not by being connected together, like the Iliad, but by dealing with one object. – What then, is it that makes man one; why is he one and not many, e.g. animal + biped, especially if there are, as some say, an animal-itself and a biped-itself? Why are not those Forms themselves the man, so that men would exist by participation not in man, nor in-one Form, but in two, animal and biped, and in general man would be not one but more than one thing, animal and biped?

This quote begins with distinctions and understandings that appear to be aligned with Koffka's notion of the whole being something other than the sum of its parts. First, Aristotle sets out the metaphysics of parts and wholes in much the same way the Koffka does: "The whole is something beside the parts." Aristotle then offers proximity (contact) and viscosity as grouping principles that create unified wholes from parts that might otherwise repose in an undifferentiated heap. *Viscosity* is not interpreted so much as a property of liquids as a generalized kind of stickiness, the sort of stickiness that creates togetherness in group formation. *Contact* does not require explanation, but it is quite interesting to see it specifically mentioned. Contact is essentially what this book is about. Aristotle is not, however, a Gestalt psychologist and the passage then takes a more philosophical/conceptual turn, asking how the concept of man may be unitary while the attributes, features, and qualities that define man are many. The issues raised in the last question are quite distinct from the issues animating this book. Here we are concerned with just the perceptual aspects of groups, and specifically with understanding how temporal flow is grouped in perception. Whether man is one thing in-itself, a unity, or two things in-itself, animal-itself and biped-itself, is not, thankfully, relevant to this undertaking.

Understanding Wholes, Parts, and Relationships

To the extent that any progress has been made in this discussion, it is that groups arise when things are seen in relation to one another. Clearly then, an understanding of what a group is will involve an understanding of what a relation is. Also clear, however, is that this is not real progress. If the concept of relation were straightforwardly analyzable, then Koffka would surely have given us a theory of grouping, and this chapter would have presented that theory instead of slowly edging up to a reckoning. The problem here is that when two things come into relation, attempting to focus on that relation and attempting to describe it in concrete and certain terms is like attempting to grab smoke. Some insight into this resistance can be gained by examining what happened to common sense and reason in the development of quantum mechanics.

Commonsense reasoning exhibits numerous forms of structure, and one structure is the rule of either/or. Applied to something like spatial position, this rule would dictate that either a thing is here, or it is there, but not both here and there. The "both here and there" structure would be an example of "both/and" thinking, and this structure is not welcome under either/or rules. What happened in physics, almost at the same time as groups were first being described in Gestalt psychology, was that the formalism of quantum mechanics required that quantum states were not "either/or" but were in fact "both/and." The technical term for this state of affairs is *superposition*. The most disturbing aspect of quantum theory is that a measurement of a quantum state is always one thing or another, adhering to the "either/or" way of thinking about the world. But to construct a mathematically coherent quantum theory, it is necessary to regard the quantum state as being "both/and" prior to measurement. So, for a particle that could be here or there, it is always measured as being either here or there, but prior to measurement it is both here and there. The world that we live in is fundamentally unknowable if we insist on knowing things through the either/or dichotomy. To fully appreciate how inadequate this mode of thought is, a person might learn about Schrödinger's "dead/alive" cat and the "both/and" interference pattern made in the double-slit experiment.

The noncommon sense presented by superposition and both/and thinking may help to explain why the Gestalt notion of group is resistant to understanding, even though groups are present in every moment of waking life. So, to the extent that we are able, let us suspend familiar either/or thinking and accept, at least provisionally, that when two things are perceived in relation to each other, they come into perceptual superposition. In this way of thinking, the experience of the group is in fact the experience of superposition between the whole and the parts as the parts come into relation with one another. There is then no definite and independent "whole" description, this state does not exist. Neither is there a definite and independent "part" description – not in experiential terms. The parts are not experienced in a state divorced from the group. Neither is the group experienced without the support of the parts. The two levels of description mutually imply each other. The superposition of mutual implying, implication running in both directions at the same time, is essentially what being in relation comes to, and what cannot be grasped by either/or thinking.

Grouping in Time

The world we live in is very busy. At any given moment much of the visual world is in motion, and this means that the photon flux arriving at the eye is carrying a complex time-dependent signal. Similarly, the soundscape surrounding our heads is filled with dozens of independent sources, each broadcasting a complex sound stream. How is it that the time-varying acoustic and photic signals flowing into our heads is not experienced as a kind of kaleidoscopic confusion of sights and sounds? This is clearly an enormous question, and it invites answers at many levels of analysis and description. A neuroscientific approach might involve a discussion of how rods and cones work, how receptive fields work, how visual cortex is organized, how the basilar membrane works, how auditory cortex is organized, and so on. Far removed from that level of description, but no less important, is Gestalt and how grouping in time works. That we live in a meaningful world is largely due to the grouping processes of temporal integration. Time-based groups carry all the subtlety and weirdness of space-based groups, and all the commentary about superposition and otherness of the whole applies equally to them. Relationship is central to what a group is, and relationship is an abstraction that is neither spatial nor temporal.

Time-based groups are experienced in a variety of ways, and a fresh look at some of the more common ones will hopefully allow us to be reminded of just how interesting our minds are. One of the most powerful experiences that life has to offer is music, and groupings in music have extraordinary properties. We will begin with music, but no less extraordinary is that we experience a world that is full of meaningful gestures, and these will also be reviewed. In this discussion it will be helpful if one takes that attitude of a tourist being introduced to some of the interesting things that people do with their minds.

Formation of Melody

Melody is good place to start a discussion of time-based groups, and a good way to introduce melody is to contrast hearing a string of musical notes with smelling a string of whatever – perfumes, fruit, vegetables. Smell is a sense that does not support the forming of relationships. The smell of an onion will not come into relationship with the smell of a carrot. There is no *onion odor–carrot odor* association in the context of smell. The smells of the onion and carrot are a good example of time-based succession. The experience of smelling an onion and then smelling a carrot is extremely literal; first there is the onion odor and then there is the carrot odor, and that is all there is. Succession is the experience of one thing happening after another where the happenings are independent and unrelated. This does not mean that odors and tastes cannot complement or conflict with each other. Conflict and complement are experiences of taste succession and are based on the way tastes chemically combine. But they do not reflect the formation of a smell-based group. Here the notion of group is exemplified by the triangle; discs as parts coming into relation to form a triangle coalition. It is this duality, having *at once* three things: onions, carrots, and the creation of an emergent onion–carrot coalition, that smell and taste cannot achieve.

The experience of a sequence of musical notes has the potential to be profoundly different from a sequence of odors. In the context of song, notes are not heard as a succession of independent and unrelated happenings. Rather, notes readily form relationships with one another, creating musical phrases and melodies. These phrases and melodies have properties that are not in the notes themselves. Melody is a strong example of a Gestalt group; the notes come together to form relations and create a melody that is other than the sum of the note parts.

Two Gestalt principles that describe the stickiness (viscosity) that brings notes together into melodies are proximity and similarity. Analogous to spatial proximity, notes that are in close temporal proximity will tend to be grouped together. A second principle is that notes that are close to each other in pitch will tend to be grouped together. These two principles may compete to produce different grouping arrangements in much the same way that color and proximity competed in Figure 2.2. Albert Bregman (1990) provided compelling examples of this competition by creating melodies that alternated high pitches with low pitches. At slow tempi, proximity determines grouping and the note sequences sound literally how they were produced; high pitches alternating with low pitches. There is, however, generally some fast tempo where the up-and-down pattern breaks apart into two separate patterns, one pattern consisting just of high pitches and one pattern consisting just of low pitches. This is an auditory example of proximity losing to similarity in the determination of group membership. Similarity also acts as a grouping principle in the dimension of timbre. Timbre is the quality of sound that distinguishes the sound of one type of instrument from another. Grouping based on timbre is quite powerful and is what allows the music played by ensembles to be perceived in terms of the contributions of different instruments. The drummer, for example, is heard independently of the bass player, who is heard independently from the melody player, and so on. All the instruments of different types (timbres) produce different streams.

Musical notes have unique relational properties and consequently the coalitions known as melodies are quite unlike any other form of time-based grouping. There is first the percept that some notes are higher than other notes. This is worth wondering about; it is a little odd that the vibrations in air that create pitch are encoded spatially. Rapid oscillations in air are perceived in a spatial way as producing a pitch that is high. It is this transformation that leads to the circumstance that a sequence of notes in a melody is perceived as a melodic contour – a melodic landscape. The spatial quality of pitch is an example of synesthesia, and it is sufficiently vivid that it dictates how music is notated. The staff onto which notes are placed is a spatial construct where high notes occupy higher places on the staff and lower notes occupy lower places. As systems of notation are nothing more than constructs, doing the opposite is a possibility – putting the high frequencies at the bottom of the staff. This is never done because the congruence between up in space and up in pitch is so strong that upside-down music notation would be bewildering.

Beyond the sense that sequences of notes go up or down in pitch is the percept of interval quality or flavor. Every musical interval has its own unique sound. For example, a C followed by an F is a fourth and that transition is immediately recognizable and can be named accurately by most musicians – it is the first transition heard in *Here Comes the Bride*. That it is the transition that carries the flavor and not the notes themselves that carry the flavor is what makes note relations unique. A fourth can be made from C followed by F or by D followed by G. All transitions that create a fourth will have the fourth flavor. It is worth mentioning that there is at least one documented person who literally tastes musical intervals (Beeli, Esslen, & Jäncke, 2005). To this person the fourth tastes (or smells) like mown grass, the minor third tastes salty, the major third tastes sweet, the minor sixth tastes like cream, and the major sixth tastes like low-fat cream.

The circumstance that musical intervals have their own special character leads to a kind of invariance in how music is heard. Because the flavor of a note interval is not tied to the notes that produce the interval, melodies are also not tied to the notes that produce them. What this means practically is that melodies can be moved around to different notes, and they will sound the "same" so long as all the note intervals are kept the same. "Same" in this context is a complicated and nuanced experience. The "sameness" arises from hearing melodies as abstract structures that are formed out of note transitions, as opposed to being formed from notes. Of course, it is not possible to play a note transition without playing notes, so notes will be required when creating a melody. In this sense, the abstract structure that is melody becomes concrete and heard when dressed in particular notes, in a particular choice of key. But because melodies are built from transitions, any melody can be dressed in any key. The melody is not in the notes, but in a sense, lives behind the notes that are heard in any given performance.

The independence of the melody perceived from the notes heard is reminiscent of how shape arises in spatial awareness. The property that objects have what we refer to as "shape" is independent of the images projected onto the retina from any point of observation. That this must be true is immediately explained by considering the consequences of the motion of our bodies or the motions of the objects surrounding us. When we move or when objects move, the images on the retina will correspondingly change. That much is just optics – ray tracing from objects through the lens and onto the retina. But we do not notice these changes because what we are perceiving are not images on the retina but object shape, and that, whatever shape is, it does not change when bodies or objects move (so long as they are not dented, flattened, broken, etc.). Somehow, we can perceive object shape in terms that are interior to the object, and in a way that effectively removes our point of view. It is as if objects are seen from all points of view at once. In this sense melodies have shape. In both cases our point of view, what is on the retina or what is in the ear, is discounted. And what emerges is an abstract percept based entirely on interior relationships.

Formation of Rhythm

One of the defining aspects of music is that it structures the flow of time. While a piece of music is playing, the passage of time is chopped up and laid out within an architectural arrangement. At the bottom of the hierarchy is what we now call the *beat* but in earlier times was referred to as the *tactus*. Tactus is an appropriate term, coming from the Latin "to touch." When our minds enter a song, the tactus touches us through the experience of music as something that is forward moving. Forward motion is not the same thing as one thing happening after another. We need a more nuanced and dynamic term and that is *pulse*. The tactus enters experience through a feeling, the feeling of rhythmic pulse. This feeling is so potent that the production and participation in rhythmic movement is part of human history and found generally in all human cultures. Gestalt psychology comes into this discussion with the recognition that the tactus, the forward motion and pulse, does not exist as a perceived quantity unless individual beats are perceived in relation to one another. Beats that are experienced as being isolated cannot form any kind of pattern and so cannot form the time grid which defines the tactus. In this way, being in relation and group formation are at the core of the feeling of rhythm. The tactus in time and the triangle in space have this much in common.

There is considerably more to the feeling of rhythm than the experience of pulse. An additional process of accenting, the labeling of some beats as strong and others as weak, gives the tactus a hierarchic structure. There are many schemes for assigning accent and each scheme is associated with a particular rhythmic feel. Accent schemes are essentially a way of creating groups within the tactus. Rhythm, in this sense, involves at least two levels of time-based grouping, and potentially many more depending on the complexity of the accent scheme. These groups are referred to in music theory as the *time signature* or, alternatively, the *meter*. Some examples will illustrate how accent is used in practice to create groups and groups within groups.

Figure 2.3 illustrates four accented groupings or meters. Each meter is displayed abstractly as an accented beat train and labelled with the time signature that would appear in musical notation. The vertical lines could be thought of as claps, drum strikes, the motion of a conductor's baton, or anything that marks a beat. Underneath the beat train are dots that are used to indicate how strongly each beat is mentally accented. More dots imply a stronger accent. The first beat is always the most strongly accented, it gets the greatest number of dots in that meter and is the listener's cue to where the grouping begins. Briefly these meters may be described:

4/4 Duple meter: The most common meter in Western music is 4/4 time. In this time signature a beat train is divided into groups of four. Typically, the first and third beat in the group of four are strongly accented while beats two and four are weakly accented.

3/4 Waltz meter: The second most common meter in Western music is 3/4. In this time signature the beat train is divided into three beats where the first beat is strongly accented and beats two and three are weak.

12/8 Meter: This is a compound time signature, also referred to as *shuffle time*, that embeds 3/4 waltz meter within 4/4 duple meter. The beat train is

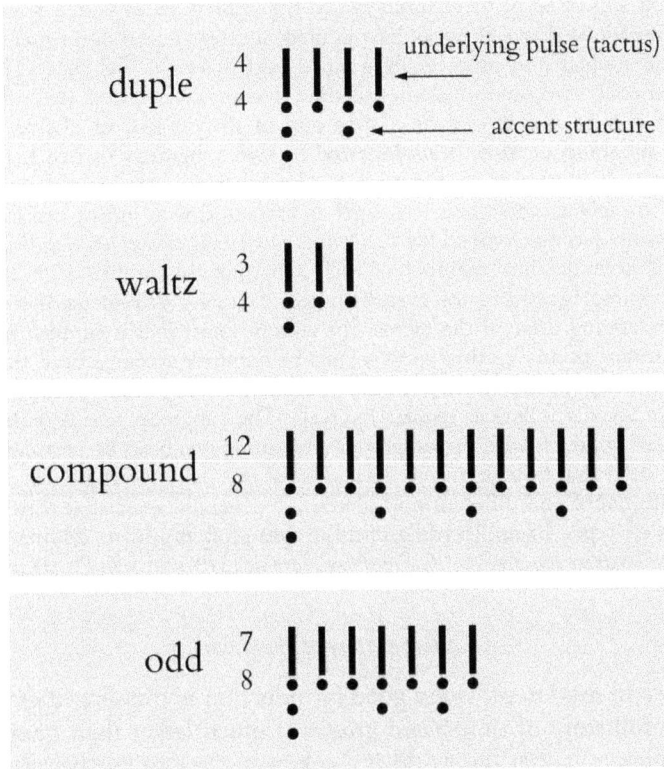

Figure 2.3

divided into groups of twelve, where each group of twelve is divided into four groups of three. Within each group of three there is a waltz feel with a strong first beat followed by two weak beats. Over the group of twelve there will be four accented beats that together create yet another level of feel: a 4/4 walking feel. What distinguishes 12/8 feel from a waltz is that the first beat of the twelve is given special accent. That accent creates the twelve group, and without it, the meter would be indistinguishable from a waltz. Consequently, in this compound meter the first beat receives three dots, and the first beat of each waltz piece gets two dots. The 12/8 meter is quite common in popular music, especially so in slow blues tunes.

Odd meters: Odd meters are common in many folk musics, especially those deriving from Eastern European, Romani, and Middle Eastern cultures. Although odd meter can be played without subdivisions, in folk music there is generally a division of the beat train into subgroups of two and three beats. These meters come in many varieties (5/4, 7/8, 9/8, 11/8, 13/8 . . .)

and almost all of them unknown in the United States with a few notable exceptions (Dave Brubeck having made a career out of odd time). One of the simpler and more easily grasped odd meters is 7/8 time. There are generally two ways music in this time signature is counted; two subgroups of two beats followed by a subgroup of three beats, or alternatively as a subgroup of three beats followed by two subgroups of two beats. The latter is illustrated in Figure 2.3. The subgroups are created by pairing a strongly accented beat (these get an extra dot) with one or two following weakly accented beats. Like 12/8 time, this meter creates a feel at the level of the accented first member of each subgroup, but unlike 12/8 time the accented beats are not evenly spaced. So, in a 7/8 meter there is the underlying drive of the tactus (the eighth notes) that is supplemented by a feeling of uneven three-ness created by the three accented beats that mark the subgroups. Odd meters have a feel that is not present in meters where the accents follow at regular intervals. The subgroup that contains three beats contrasts with the subgroups containing two beats by creating a sense of delay that feels expansive. That sense of expansiveness translates naturally into dance and different odd meters are generally associated with specific dance types. Examples of Eastern European folk music in odd meter music are *Jovanne Jovanke* in 7/8, *Gankino Horo* in 11/8, and *Sedi Donka* in 25/16.

Choreography of Behavior

Melody and rhythm provide a good introduction to time-based grouping, but the full story of time-based groups is much larger than music. The bigger picture is that the world is a coherent place to live because time-based grouping is continuously presenting us with meaningful content. At a physical level the world presents itself as an ensemble of surfaces and edges of which quite a few are making noises and moving about. All that activity would be kaleidoscopic if it were not organized. While spatial organization creates objects, temporal organization creates events and scenes, a procession of "this" happening and "that" happening. The structures referred to by the "this" and the "that" are groups, packages of temporal flux that have been glued together to make an event that has some level of ecological significance. In this discussion it will be helpful to have Koffka in mind. Although Koffka was principally interested in spatial forms of grouping, his comments apply equally to the temporal packages that constitute the experienced world; events are "other" than the sum of their temporally distributed parts. Instructive examples of the "otherness" of temporal packages are to be found in ordinary human activity.

In the arena of human activity, there is a choreography of body motion that proceeds and is structured through the formation of kinematic groups.

An example from the world of gestures will make clear just how ordinary this process is. Consider the shake of the head up and down that indicates the meaning of "yes." The head nod consists of three separate levels of structure. There are first the individual head motions. These individual motions are like the notes of a melody or the vertices of a triangle in that they are the parts that form a whole – something "other," when they come into relation to one another. When the individual up-and-down motions are perceived to be in relation to each other, both by the actor and by the person that is receiving the nod, they group together and something like a melody or shape emerges. Specifically, what emerges is a gesture that is rich with the meaning of "yes" (at least in some places). The third level of structure is the supervisory awareness that nods have a moment in which to live. There is an overarching sense of when the nodding has proceeded to the formation a group that suffices to create the gesture. This is a situation where more may not be better. A person that is nodding their head for more than a few seconds is not communicating assent. A prolonged nodding episode could be communicating sarcasm or perhaps a medical condition.

The grouping of repetitive motions is not limited to communicative gestures and extends to virtually all things that people do with their bodies. The formation of action-based motion groups includes ordinary human activities such as cooking (e.g., chopping, stirring, rubbing), working with one's hands (e.g., hammering, sawing, sanding), and grooming (e.g., brushing hair, scratching, exploring ears and nose). In all these cases individual hand and body motions are packaged together to form meaningful events. These packages, unlike gesture, do not serve a communication function but they are groups, nevertheless. The stirring of a pot, for example, typically involves several circular motions with a ladle. These motions are not perceived by the actor as being isolated events, but rather they are perceived in relation to one another. The activity of stirring is the "otherness" that is created when a sequence of circular motions forms a group. "Stirring" is not present in any circular motion, it arises at the level of the group. Recognizing that cooking motions like stirring do not have a communicative function, it is of some interest to inquire what kind of supervisory processes regulate the length of stirring and other noncommunicative motion groups. To answer this question, it would be necessary to first film people while they are cooking, working, grooming, and doing whatever else they do, and then to measure the durations of motion sequences. This has been done by Margaret Schleidt and her colleagues (Feldhütter, Schleidt, & Eibl-Eibesfeldt, 1990).

The research focus of Schleidt and her colleagues was not on temporal grouping as something of interest in itself, as it is here, but rather specifically on the durations of motion-defined groups. The entirety of their work is encapsulated by the duration histograms they produced for both grooming and working motions across a range of indigenous peoples observed (Himba, Trobiander, Yanomami). These histograms were produced by simply counting how many times a motion-defined group consumed a given amount of time. A deep analysis of the duration histograms was not required or offered because it was obvious that they were substantially peaked in the two to three second range. That is, motion-defined groups typically last for just a few seconds. Schleidt and her colleagues did not speculate on why this is so, but that it is so may be an important discovery. There is something intrinsic to the way people choreograph their behavior so that across cultures and across behavior types the duration of two to three seconds is recurrent. Schleidt and her colleagues explicitly point to this recurrence in the title of one of their articles: "A universal constant in temporal segmentation of human short-term behavior" (Schleidt, Eibl-Eibesfeldt, & Pöppel, 1987). It is important to underscore that this discovery does not require specialized equipment or even a laboratory. Sensitive and aware appraisal of motion-defined grouping is something that anybody can practice at any time.

Perspective on Grouping

Understanding the basics of perceptual organization essentially comes down to grappling with two ideas. The first is that when things come into relation with one another, groups are formed. The second idea is that groups are "other," that there are properties flowing out of the relations that belong to the group alone, properties that otherwise would not exist. The apparent simplicity of these ideas should not distract from how important they are. Groups are the medium through which an animal mind creates a reality that transcends the physical world. Difficult to understand but essential for understanding Gestalt is that the physical stuff of the world does not appear in experience. What appears is an environment that is built out of relationships. From these relationships emerge the objects and events that compose what we call reality. The world that we experience is an organized world, and as such it is built upon a myriad of forms of "otherness." Or we could just say the world is other because there is no world for an animal other than the experienced world.

It is from this vantage point of relationship that we can begin to build a theory of human temporality. This theory begins by noticing something that is quite interesting, that there are circumstances that defeat the construction of time-based groups, where events do not come into relation, and are not put together. To the extent that the perceived world is built out from relations, the most important thing nature gives us is time to create these relations. What happens when time runs out? This question will lead to new perspectives on how people live in time.

CHAPTER 3

Near and Far in Space and Time

Groups in space share many properties in common with groups in time. Both group types come into being when things are perceived as being in relation to one another. Also, both types of groups are "other" than the sum of their parts; groups always have emergent properties. And in some ways proximity in time acts as a grouping principle in the same way as proximity in space. Being a neighbor generally leads to being perceived as being in relation. There is, however, one critical difference between proximity in space and proximity in time, and that is how groups behave when they are dilated or expanded.

Proximity in Space

In some respects, distance is determinative of how spatial groups are formed, and in other respects distance is completely unimportant in grouping. How distances can have both properties is explained by how spatial intervals in the world are projected onto the retina. Visual process begins with a lens that focuses light onto the retina. Although the visual system is not a camera, the front end is very camera-like, and camera optics are relevant to the perception of spatial groups. The way light starts out as reflections from surfaces and ends up as retinal stimulation explains how spatial groups may be both exquisitely sensitive to spatial separation and yet not be affected by uniform expansion. Viewing a group and its expansion will clarify this duality.

In Figure 3.1 a simple grouping arrangement is shown in both large and small versions. In both versions the percept is of two quartets of discs, one on top of the other. The way the quartets group in the smaller version does not appear to be different than the way the quartets group in the larger version, even though the discs in the smaller version are closer to each other. Admittedly, there is some fuzziness here in referring to "the way quartets group." Although the language may be imprecise, the two

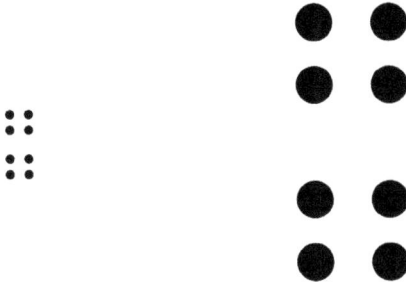

Figure 3.1

arrangements do "look" the same except for size. The quartet groups in the two versions look the same because all the distances and sizes have been multiplied by the same factor to make the larger version from the smaller version. Specifically, the distance between the discs relative to the distance between the quartets is the same in both versions, as is the disc size relative to the distances between discs. The "looking the same" is not a trivial observation insofar as it reveals something basic and important about spatial vision and spatial grouping. The implication of "looking the same" is that as far as grouping goes, proximity in space is relative. Distances, in other words, are not perceived in isolation, but as contextualized in terms of whatever other distances may be present. Another way to say this is that distances are perceived as contrasts, or in a more mathematical language, as ratios. In this sense, both the distance between discs as well as the disc sizes are effectively divided by the distance between quartets to arrive at an impression of what is close and what is not close. It is the "dividing by" that cancels out the dimensions in spatial vision and makes the perception of spatial relations dimensionless. Whether quantities are dimensioned is not typically an issue in psychology, but here it is quite important. And the first thing to be pointed out is that the visual system works with ratios that are dimensionless quantities.

The ratio nature of visual scaling can be traced back to the way light is projected onto the retina by objects in the world. Both with cameras and with eyes, the size of the image cast onto the film plane/retina depends both on the size of the image-casting object and how far away that object is from the point of observation. In the language of optics, every object subtends an angle from the point of observation, and that angle will be large or small depending on both the size of the object and its distance from the eye. The situation is illustrated in Figure 3.2, which shows four versions of the quartet

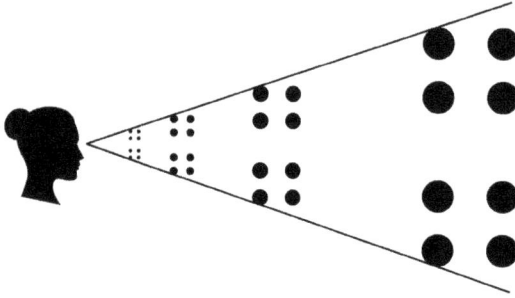

Figure 3.2

groups in a cartoon that demonstrates how they might all project onto just one retinal image. All versions have the same internal ratios and have been placed so that they fit exactly into the same subtended angle. In this way the cartoon suggests that they project the same image onto the retina. Given just the image on the retina, there is no way to recover which of the four versions is the quartet pair that is in the world producing the image. Size does not matter when it can be traded off with distance from the eye.

Size not mattering in the formation of groups is obvious but important. That groupings in postcards do not look different from groupings in the 3D world is kind of profound when it is appreciated just how much larger the world is than a postcard. What the irrelevance of absolute size in the grouping process means is that there cannot be a discovery that, say, objects further apart than two feet will fail to form a group. This statement may seem like a truism, something so patently true that it does not need to be said, but it will have a special meaning when the discussion turns to time-based groups.

Proximity in Time

Size may not matter for spatial groupings, but it matters quite a bit for time-based groupings. This becomes clear when we observe what happens when a time-based group is expanded so that the various occurrences that form the group become more separated. We know what expansion does to space-based groups; proximity is relative and so group assignments are not changed by expansions. What expansion does to time-based groups is more complex because the experience of *nearby* in time has considerably more structure than the experience of *nearby* in space. A few examples will illustrate what happens when events become increasingly spread out in time.

Expanding a Melody

Expansions in the context of music are nothing more than a slowing of tempo. As the tempo slows, notes are held for longer and the time intervals between notes, rests, expand in exact proportion as well. Perceptually, there is little difference between a song played at 120 bpm and a song played at 100 bpm. The slower version, of course, sounds slower, but because a tempo shift preserves all the ratios between note lengths and rests, the song sounds like the same song at both tempi – at least to the extent that the song is recognizable independent of the tempo. In this regime, expansion in time behaves exactly as expansions in space with preservation of ratios being the key determinant in the perception of sameness. That songs sound the same when played at different tempi will be obvious to musicians insofar as it is inevitable that choices are made about how slow or fast to play any given piece of music. These choices are only possible because music does in fact sound the same when played at different tempi. But the experience of sameness with tempo change is true only up to a point. And that point is the focus of this chapter and the theory of temporality that follows.

There is always a tempo that is so slow that a melody drags to the point of being unrecognizable. At a sufficiently slow tempo, the notes stop being heard in relation to another. That does not mean that perceptual organization in time has ceased. Rather, what happens is that individual notes disconnect from their neighbors and become islands of sound that appear one after another, in succession. What is lost when melody becomes note succession? In a word, everything – everything that is created when notes form a group. When notes disconnect and the group dissolves there is no contour, no note intervals with their special flavors, and no transposition. Still, a succession of sound islands creates a distinctive experience because a sound island is not free-floating, inhabiting a vacuum of nothingness. A sound island is surrounded by silence, and it gives that silence structure. It breaks up the silence and the silence is heard as if it is interrupted. The silence accretes around each disconnected note creating individual beginnings and endings: Each note has its moment. The mind never stops organizing, and when the notes can no longer group with each other, they will group with the silence that surrounds them. Nothing in human experience escapes organization.

In a study that appears to be unique in the literature, Warren et al. (1991) measured the transition tempo where melodies become a succession of sound islands. The songs in this study were familiar songs such as *Happy*

Birthday and *Twinkle, Twinkle, Little Star*. The logic of the study required that the songs be played in a slightly artificial way with all notes having the same length. Only then would the spacing between notes be uniform. Warren et al. accomplished this by using MIDI instruments and assigning notes a common numerical duration. Between notes a brief gap of 20 ms was placed so that repeated notes were heard as repeated. So, in this experiment tempo translates into the duration that each note was given. A tempo of 120 beats per minute (bpm) would translate to each note having a duration of about 0.5 s. The procedure in this experiment required only that participants attempt to identify songs after listening to a small snippet. For every song there were note durations (tempi) where it could be identified and named, and durations where identification was incorrect or not attempted. The aspect of the data most of interest here is the transition duration at which the songs began to dissolve and identification became difficult. Warren et al. found that over participants and songs the median transition duration was about 1.3 s or equivalently at a tempo of about 45 bpm. As we shall see below in further discussions of group dissolution, this is a highly credible estimate.

Expanding Rhythmic Pulse

The feeling of rhythmic pulse has the same dilation phenomenology as the experience of melody. There is a regime of tempi where rhythmic pulse is felt, but there is also a limit. At sufficiently slow tempo, the feeling of rhythm is not possible. Again, this is obvious, as a song that is played at one beat per day will be difficult to perform without a clock. But just exactly how slow that tempo must be to dissolve the feeling of rhythm is one quantity that is quite well known. The reason for this is that there are manufactured objects that are used by musicians to help them play exactly in time. These objects, metronomes, were originally mechanical and made of out of real material – metal and wood. Wood and metal metronomes have a wand that swings back and forth, and its period is set by the position of a movable weight. The wand and weight system form a pendulum, and the physics of the pendulum is used to set the tempo. The pendulum has a natural period that is proportion to the square root of its length. Length in the metronome context means how far out the weight is from the fulcrum, and so it is inherent in the fabrication of these metronomes that long wands must be installed to reach slow tempi. From a manufacturer's perspective, the size of the wand sets the size of the wooden case, and ultimately the cost to build the metronome. Economy dictates that the design of the metronome reflects realistic tempo constraints on the experience of rhythmic pulse. There is no reason to make a metronome that can provide evenly spaced clicks at tempi that do not afford the feeling of rhythm. In this regard it is telling that the slowest tempo that manufactured metronomes support is 40 bpm. This is an interesting number. It may not represent the tempo where people are unable to experience rhythmic pulse, but it is cautionary; do not attempt to play music slower than 40 bpm.

Forty beats per minute correspond to an interval of 1.5 s between each metronome click; 1.5 s of separation for the beginning of pulse disintegration is in close agreement with the time interval between note onsets where melody disintegrates. A unified picture of musical groupings seems to be emerging. At time separations less than about 1.5 s both notes and beats form groups. Beats coalesce into a tactus, which is organized by an accent pattern into a meter. Notes coalesce to form transposable melodies that are experienced spatially as contour. And at time separations greater than about 1.5 s beats and notes dissolve into successions of unrelated events.

These commonalities suggest a general template for understanding time-based grouping: When events are sufficiently proximal in time, they may be perceived in relation to one another. These relations lead to a myriad of emergent properties that form the basis of ordinary world experience. The regime of grouping and emergence defines one phase of temporal experience. When events are not proximal, when they are too separated in time, they are perceived as one thing happening after another in succession. The regime of succession defines a second phase of temporal experience. Separating these two phases of experience is a transition point, a *phase transition*.

It is the existence of phases and a phase transition that makes temporal grouping radically different from spatial grouping. In spatial grouping there are no phases defined by proximity, distances are relational, and there is no role for an absolute unit of distance. But in temporal grouping there are phases, and there is a time separation that is not relational, it is absolute and to the extent that the template is universal, it is about 2 s. The centrality of this distinction cannot be overstated. In spatial perception there are no units, there are angles subtended by objects, but no inches and no feet. However, in time perception there are units, time is measured in seconds, and 2 s is apparently when grouping ends and succession begins.

Ethology of Group Dissolution

Although already said, it is worth saying again: Everyday life is not experienced as a confusion of color, movement, and sound. Rather, the shifting landscape of sight, sound, and touch is packaged into time-based groups – events or happenings. Of the many things that humans and other animals are good at, they are especially good at creating and recognizing events within the scope of their ecology. One consequence of this knowledge is that people have a tacit understanding of how time-based groups work and how to manipulate events so that they fall in or out of groups according to what may be needed or desired. So far, we have some evidence that time-based groups are not particularly elastic and that they will dissolve if parts are stretched out beyond a couple of seconds. If this limit has generality, then it should be recurrent in the decisions that people make as they evaluate and manipulate time-based groups. This is an opportunity to do some ethology, and there is no question that the most fun way to do an ethology of group making/breaking is to delve into the methods sections of academic psychology articles. In the methods sections of garden-variety experiments, ordinary people (psychologists) reveal how they think about temporal group formation through the way they pace the delivery of stimuli and trials. One of the interesting things that will be learned here is that timing decisions are rarely justified in methods sections. They are not justified because these articles are written by people for people. Because people implicitly understand how temporal grouping works, and what its limitations are, no justification is required.

How to Conduct a Memory Study

Since Ebbinghaus conducted the first formal studies of forgetting, memory studies have largely involved the learning of lists. One reason that lists are ideal for the study of forgetting is that a forgetting score is easily assigned. Lists consist of items, and these items are either remembered or they are not, and this binary outcome is easily scored. A second reason is that lists are inherently forgettable, a property that permits forgetting processes to be observed on timescales of tens of minutes. List learning raises the procedural issue of how lists might be delivered by experimenters so that they can be learned by participants. The procedures that experimental psychologists adopt for list delivery provide an opportunity to do some ethology on peoples' informal understandings of group formation and dissolution in time.

A list is distinguished from other forms of discourse by its grouping, or more exactly, its lack of grouping. It is imperative in the reading and reception of a list that list words are not heard in relation to each other, that they are heard as single and unitary constructs. In practice, list words are detached from each other by the insertion of pauses. A word surrounded by a substantial pause will detach from its neighbors and come to rest in a pool of silence. In this respect detached words in a list are like notes detached from a melody. The practical issue of how slowly music must be played to create note detachment arises here as the question of how slowly word lists must be read to create word islands. List words could be delivered at one per day or one per hour if detachment were the only issue. But these are not the separations encountered in the methods sections of memory articles. As an example, consider an influential article on remembering words in lists that was designed around the concept of false memory (Roediger & McDermott, 1995). The authors, having a wide and lifelong experience of time-based groups, know exactly the minimum rate at which words might be read so that they are experienced as being detached from one another. In their methods section, it is casually mentioned that words were read at the rate of one word every 1.5 s. This much time suffices to create word islands, both in the minds of the experimenters, as well as in the minds of the participants. The choice of 1.5 s, again, is mentioned but not justified. It is the absence of justification that makes this valuable ethological data.

How to Conduct a Lexical Decision Study

Cognitive theory is largely based on mental chronometry, the careful measurement of mental processing times. The hope and purpose of this form of measurement is that patterns in the speed of response might specify how the mind operates and how it is structured. The most common methodology in mental chronometry involves people deciding, as fast as they can, whether an image is an *A* or a *B*. *A* or *B* could be whether a target is present or not in a visual search study, whether a figure has been mirror-inverted or not in a mental rotation study, whether a letter string is a word or not in a lexical decision study, and so on. In every case the participant is instructed to answer as quickly and accurately as possible. The two quantities measured are the response time latency to make the *A* or *B* decision and whether the decision was correct or not.

Time-based grouping comes into this picture through the circumstance that response time latencies are highly variable. People might consistently come to the exact same conclusion upon multiple viewings of a given display, but the times to arrive at this conclusion will not be the same. This is a natural outcome of being alive; our nervous systems are warm and in a continuous state of fluctuation in readiness and responsiveness. Although it might seem counterintuitive to regard response times as random numbers, response time latencies are typically understood as probabilistic quantities, selections from a probability distribution. There is, to be sure, some part of a response time that is not random, that is systematically related to the *A* versus *B* difference the participant experiences, but there is no question that the time it takes to generate a speeded decision of any type contains a large random component. The goal, in fact the single goal, of most experimental psychology is to sift through the randomness to find the part that is systematic. In typical experimental methodologies this is accomplished by presenting many *A* versus *B* decisions, each decision in a separate trial, and the sifting occurs through the computation of response time averages. And it is in the experience of the myriad trials that grouping becomes an issue for the participant.

There is no manual for how to do experimental psychology, and so it is inevitable that every laboratory will create its own practices and conventions. A common problem faced by every laboratory, although never acknowledged in the literature, is figuring out how to construct experiments that are doable, that make sense to participants and potentially lead to meaningful data. At the core of this problem is figuring out how to create trials that make sense, that are experienced as complete and whole with definite beginnings and endings. Trials that are experienced as complete and whole are trials that are experienced not in relation to each other, but in succession – one after another. The ethology of how trials are created to be in succession turns out to be a clinic in the segmentation of time-based groups.

The example chosen for illustration here is taken from Meyer and Schvaneveldt (1971), one of the most influential articles in the history of cognitive psychology. In this article, the task, referred to as *lexical decision*, involved deciding whether pairs of letter strings were both words, or if one of the letter strings was a nonword. The rationale behind the lexical decision task was to expose a phenomenon known as semantic priming. In a later chapter semantic priming will be discussed in terms of its time course, but here the focus is just on how the trials were constructed. In this pioneering study, the trial structure was complex, consisting of a series of discrete episodes. Trials began with the brief appearance of the word "ready." This was followed by a fixation box that was displayed for 1 s. A fixation box is simply a box where people are encouraged to look so that they can see what happens next. After these two preparatory events, the letter strings finally appeared. The participant then responded, hopefully as quickly and accurately as they could. Following the response, there was a 2 s period during which the participant was given feedback about the correctness of response.

The durations of the various episodes employed in these trials are meaningful and they are instructive in how people experience time passage. These trials began with what is essentially an instruction for the participant to focus their mind. The 1 s fixation box between "ready" and the letter strings does not create a segmentation. That is, "ready" is not disconnected from the letter strings. The word "ready" creates a focused mind, a bridge to the stimulus that will be acted upon. Why does the fixation box not interrupt for 1.5 or 2 s? Because these authors do not desire for "ready" and the subsequent letter strings to be experienced as separate events, as islands. They want "ready" to create a context into which the letter strings are dropped. Extending the fixation box time out beyond 1 s invites not a mind that is prepared, but a mind that is wondering what is going on, a mind that is wandering.

There is a connection to be made here between fixation boxes and ballads. Ballads are a distinct form of musical style that find their identity in exuding a kind of languorousness and longing. They are typically played at a tempo of about 60 bpm and often even slower at 50 bpm. The languid feel of a ballad is ultimately tied to the bridge building that may occur across gaps of about a second. When notes are separated by an average delay of a second, there is the emergent sense that each note lands into a pool that was created by the note preceding. Although Meyer and Schvaneveldt may not have been thinking about ballad structure when creating their fixation box, they are using the temporality of the ballad to create the anticipated moment when a letter string lands in a mind prepared by "ready."

The connection between music and lexical decision methodology continues with the choice of a 2 s separation between trials. A 2 s gap does not create a bridge; it creates a moat – a separation. The phenomenology of a 2 s gap in trial sequencing is that it allows the trial to be apprehended as having ended in a state of completion before the next "ready" signal appeared. The same 2 s gap in a note sequence would have the same effect. Notes would detach from their melodic context and come to be experienced as complete in themselves. The general situation in music is that musical phrases are offered as complete experiences, not as collections of individual notes, making 2 s gaps undesirable. But in an experimental design, it is not desired that the trials make music. The trials must be experienced as separate, as in the sense of one trial happening after another. Essentially that is why these authors "play" their experiment at 30 bpm.

Persecution by a Number

A meta-observation about experimental psychology: Psychology experiments generally do not lead to the production of numbers that need to be taken seriously. More to the point, experiments that are successful generally produce meaningful data without producing numbers that are meaningful in themselves. Replications or extensions of well-known psychological effects always produce different numbers even when the interpretation of the numbers leads to the same conclusions. The issue is not just statistical variability, which always exists, but rather that the numbers received in an experiment inevitably depend upon background human factors such as the acumen, taste, and experience of the investigators. These human factors will shape decisions about experimental design, stimulus construction, procedure, and task composition. Ultimately, there is no one canonical way to compose and conduct an experiment and this means that what experiments produce are effects, not numbers. This circumstance is tacitly acknowledged by the way data are analyzed in psychology. The question asked in a typical statistical analysis is whether the treatment condition leads to *different* data than the control condition. The specific data values are rarely the issue. This is the general situation in experimental psychology – the production of meaningful effects and trends built upon numbers that could have been other numbers and will be other numbers when the experiment is replicated or extended.

Yet, there was a moment in the history of experimental psychology when a meaningful number was encountered. George Miller, writing at the dawn of cognitive psychology (Miller, 1956), describes how he is persecuted by an integer. He writes:

> My problem is that I have been persecuted by an integer. For seven years this number has followed me around, has intruded in my most private data, and has assaulted me from the pages of our most public journals. This number assumes a variety of disguises, being sometimes a little larger and sometimes a little smaller than usual, but never changing so much as to be unrecognizable. The persistence with which this number plagues me is far more than a random accident.

This integer is 7, written by Miller as 7±2 to indicate that the 7 has disguises.

The number 7 arose as an observation within a series of experiments that Miller and a group of experimental psychologists ran in the 1950s. What they were measuring was a cognitive limit in absolute identification. An example of absolute identification is knowing the names of people you recognize. The name absolutely identifies them. Miller and his cohort were not interested in something as complex as a face, but rather in identifications within a single continuum of experience – such as loudness, brightness, size, and generally the continua that Stevens would have placed within his law. So, in their experiments people had to name specific pitches, specific levels of loudness, specific levels of salinity, and so on. Going into these experiments Miller and his cohort knew that naming a single thing when it is placed in front of you is much more difficult than deciding whether two things are the same or different. For example, when looking at a black and white picture it is evident that there are hundreds of discriminable shades of gray. But can you imagine assigning numbers to, say, just 20 or 30 shades of gray and then correctly assigning their number when presented each shade one at a time? It is here that the persecution begins. The number of shades of gray, for example, that a person can keep straight in the sense of being able to name them is in fact about 7 (the zone system in black and white photography uses about 10). In his article Miller presents a variety of naming studies in a rhetorically effective argument that 7 is a recurrent limit to what can be identified with a name. In these studies, it mattered little what kind of experience was being named or the details of how the levels were selected. This is not the expected outcome when doing psychology. The expected outcome is that while there may be robust effects, the numbers, the data, will always reflect the way the experiments were constructed. A number that will not go away might well be experienced as persecution.

Persecution by Another Number

Grouping is inherently an extremely abstract topic. It is built upon the concept of being in relation, which is nonanalytic and so quite resistant to coherent description. Being in relation inevitably leads to the experience of groups being-other, another nonanalytic concept that expresses the truly magical quality of everyday experience. Yet, despite the mysteriousness of the grouping phenomenon, there is the appearance of a recurrent quantity. There is the recurrent observation that time-based groups disassemble into event succession when the temporal proximity of neighboring events exceeds a couple of seconds. Like the magic number 7, the group killing span is not a quantity that will be exactly 2 s in every case. Sometimes it will be observed to be a little smaller, sometimes a little larger. But never so different from 2 s that it would be unrecognizable. Inspired by Miller, we shall refer to this span as 2±1 s, and from here on out it shall be recognized to be what it is – a proximity constraint on grouping; 2±1 s creates a meaning for proximity, what counts as a long wait, and defines a critical phase transition in human experience.

Everything in this book flows from the single observation that there are two phases to temporal experience. On the short side of 2±1 s, when events arrive close enough in time to form a group, we experience all the ways in which groups are "other" than the sum of their parts. This is a large category as it includes all the emergent properties of groups that make life meaningful and interesting. In contrast, on the long side of 2±1 s we experience a wasteland where the varieties of "otherness" are reduced to the sameness of succession. A melody, for example, that is played so slowly that the individual notes have detached from one another will lose its uniqueness and individuation. It will not sound that different from any other detached note sequence. In this sense all detached melodies form a class – the class of meaningless detached-note music. Similarly, a drumming performance that is wandering because the experience of tactus, of pulse, is not available at a slow tempo, will not sound that different from any other wandering and lost performance. This is another class – the class of lost pulse and incoherent drumming. The importance of this phase transition should now be clear. On one side is meaning and the diverse forms that meaning takes. On the other side is a uniformity produced by the meaninglessness of one-thing-after-another. When an event is detached and surrounded by silence it makes little difference what the event is; it is experienced as an island.

In this chapter an effort has been made to demonstrate that the span of 2±1 s is a recurrent quantity in the phenomenology of grouping. Most of what has been presented has been based on informal observation and ethology. Informality in this context is not a sin. This is an informal enterprise because everything we need to know is literally lying on the ground; it is the lowest of hanging fruit. As this fruit is littering the landscape of everyday experience, it has not escaped the notice of other psychologists interested in the limits of time-based grouping. Of particular interest to the inquiry here is how other researchers have framed the notion of phase transition and what they believe counts as evidentiary support for the destructive power of 2±1 s.

Paul Fraisse, "bit" and "ter," and the Capacity of Apprehension

It is not possible to delve very deep into the history of psychology without coming into to contact with the work of Paul Fraisse. His definition of the psychological present is foundational in the field of human timing. He writes (Fraisse, 1978):

> The perceived present, or psychological present, may be defined as the temporal extent of stimulations that can be perceived at a given time, without the intervention of rehearsal during or after the stimulation. The capacity of apprehension of successive stimulations is to be distinguished from long-term and short-term memory.

The psychological present is one more exceedingly abstract concept and consequently there will be different points of entry in efforts to give it definition. Fraisse does not use the language of being perceived in relation, preferring the language of "perceived at a given time," but they mean the same thing. In this passage Fraisse gives the notion of the subjective present some specific content by connecting it to grouping. Fraisse effectively defines the psychological present in terms of a proximity constraint on group formation; the psychological present is the span of time over which separate events may appear to be in relation to one another. The "capacity of apprehension of successive stimulations" refers to a memory system that permits events occurring at different times to be perceived "at a given time" – to be perceived in relation. This memory system is not identified by Fraisse except through what it is not; it is neither long- nor short-term memory. The positive identification of this memory system will take up the remaining chapters of this book.

Fraisse also offers some empirical commentary on the capacity of apprehension. This offering is given not in the form of experiments and data, but in the form of personal introspection. Fraisse wonders how much time can be inserted between *tick* and *tock* and still preserve their sense of going together, how much time can be inserted between two notes and still retain their quality as a musical interval, and how much time can be inserted between *bit* and *ter* and still retain their fusion into the word *bitter*. We learn that in each case the maximum interval is around 1.5–2 s. From the perspective that is being developed here, what Fraisse is divining is the position of the phase transition between group formation and succession. That he converges on the value of 2±1 s is interesting, but not as interesting as the informality with which these estimates are offered. In so far as there is no reference made to published research on any of these phenomena, it is evident that Fraisse does not believe that specific citations are required to support his claims. The conclusion must be that Fraisse regards the destructive force of 2±1 s to be common knowledge. If there are critics of the idea that 2±1 s is a phase transition between grouping and succession, Fraisse does not acknowledge them. There is, perhaps, no greater testimony to the "everybody knows these things" than Fraisse's choice of the word *bitter* and the sound of *tick-tock*. Where do these instances come from? The utter arbitrariness of these selections is implicit testimony that, really, examples of 2±1 s are literally everywhere.

CHAPTER 4

Memory With and Without Remembering

The discussion so far has been little more than a primer on the most elementary aspects of everyday experience; that there is grouping in both space and time, and that there is a phase transition in time-based grouping at 2±1 s that separates perceiving in relation from perceiving in succession. All of this is common knowledge. This knowledge is present in the metronome, it is present in the methods sections of academic psychology articles, and it is apparently so pervasive that Fraisse only has to mention the sounds of *tick-tock* and *bitter* for his concept of the capacity of apprehension to be clear. That something is common knowledge does not, however, devalue it. The existence of a phase transition in time has implications for the nature of human temporality, implications that are far from common knowledge.

At face value, the mere existence of time-based groups implies the activity of a memory system. In the most elementary terms, the time course of mental life comes down to: One thing happens and then another thing happens, and under the right circumstances these two things are perceived to be in relation with one another. If so, they form a group, the group is "other" – it has emergent properties, and these properties make up the experience of the world. An organism that is truly amnesiac would not be able to bring two events separated in time into a relation. Every event would be experienced as if it were the first event. So, we can conclude that some kind of memory is involved. However, forgetting in the context of grouping is not like forgetting a name or where the keys are. Attention must be paid to how a phase transition operates to appreciate how memory operates in a grouping process.

The first step toward understanding what form of memory is involved in grouping is to clearly identify what is forgotten when a memory system that supports group formation fails. Fraisse's example of *bitter* will serve here. When the time interval between the two syllables, *bit* and *ter*, is short, less than a couple of seconds, the two syllables are heard in relation to one another and they form a unity, a word. When the time interval between the two syllables is longer than a couple of seconds, something is lost as *ter* detaches from *bit* and becomes just a sound that happened after the sound of *bit*. What is not lost are the syllable episodes. Having heard *bit* followed a while later by *ter*, there may well be recollection of the two syllables, that they were spoken and that they were heard. The memory of that episode may persist for years or a lifetime. What is lost when events enter the phase of succession is the connection, the being in relation, the togetherness, the glue that binds the two syllables into a word. Evidently, forgetting in the context of grouping involves not a loss of the parts of a group, but a loss of the sense of the togetherness in a group. Losing the sense of togetherness is a very different kind of forgetting from blanking on a name or losing the keys. Understanding it will require significantly enlarging the concept of memory.

A difficult and somewhat abstract issue arises here in contemplating what a togetherness memory might involve. First, it is obvious that different groups have different characters. The binding of *bit* and *ter* into a word has a different flavor than the binding of notes into a melody, which has a different flavor than the binding of beats into rhythmic pulse. In these three examples the parts are different, the groups are different, and the "otherness" – the emergent properties – are different. Yet there is a commonality in that grouping always involves a sense of togetherness, of being in relation. There also seems to be commonality in the unrelating that leads to the phase of succession. There are surely no data on this, but it does seem that the pool of silence that surrounds an isolated word in a study list is not fundamentally different from the pool of silence that surrounds an isolated beat or note. And perhaps most importantly, 2 ± 1 s has emerged as a recurrent and persecuting constraint on proximity. So, although there are differences and variety in what emerges from group formation, there is, nevertheless, a convergence in the ways groups unbind into succession. This convergence might imply that there is a unitary memory system that underlies and supports the formation of relations and groups. That certainly would provide a concise account for the recurrence of 2 ± 1 s in the group to succession transition; *one* memory system, *one* span, *one* phase transition. In this regard, it is noteworthy that Fraisse mentions a single capacity of apprehension that underlies a single subjective present. He does not regard the unbinding of *tick* from *tock* as having a different flavor or perceptual quality than the unbinding of *bit* from *ter*. The alternative to a unitary memory system is that there are, perhaps, as many memory systems as there are forms of groups. In view of the circumstance that life is full of new experiences, and that those new experiences inevitably involve new groups, new "otherness" – new emergent properties – and new forms of successions, it is difficult to imagine the spontaneous creation of new memory systems to support these new groups. In the absence of evidence to the contrary, we shall refer to grouping or togetherness memory as just one thing.

There is no theory within cognitive psychology of togetherness memory, the *capacity of apprehension* that underlies the percept of being in relation and togetherness. Apparently, Fraisse has no theory of this capacity even though he can elicit examples with the certain knowledge that his readers are in complete agreement. Nevertheless, Fraisse is clear about what the capacity of apprehension is not – it is neither short- nor long-term memory. It clearly is not long-term memory insofar as the capacity is limited to 2±1 s. The *long* in long-term memory refers to time spans greater than seconds. But that Fraisse distinguishes the capacity of apprehension from short-term memory is highly instructive. The conception of short-term memory that existed at the time of Fraisse's writing, and which continues to the present day, is ill-suited to understanding the kind of forgetting that forgets relatedness and which unbinds groups. This conception is, nevertheless, quite valuable as a negative foil for thinking about memory systems that do have the appropriate structure for group formation and dissolution. The story of short-term memory, or *working memory* as it is more commonly known today, begins in the 1950s, but as the story has elements that are more than a little odd, it will pay to start the story some 60 years earlier with the originating studies of forgetting conducted by Hermann Ebbinghaus.

Ebbinghaus

Any inquiry into the nature of forgetting will eventually lead to Hermann Ebbinghaus, a German researcher who was instrumental in transforming psychology into an experimental science. Ebbinghaus was part of a larger movement centered in Germany and Austria that sought to bring scientific method and systematicity to the descriptions of psychological phenomena. Systematicity in a psychological context did not imply equations of motion derived from first principles, but it did lead to the development of a range of quantitative techniques for the study of sensation and perception. With quantitative techniques came various laws of perception, Weber's law and Stevens' law (see Chapter 1) being two notable examples. The phenomenon of forgetting was brought into this intellectual mix when Ebbinghaus produced the first mathematical relations between memory retention and time since learning (Ebbinghaus, 1948).

The kind of memory that Ebbinghaus studied would be referred to as long-term memory, simply because the retention intervals used in the paradigm he invented spanned from minutes to days. There are many forms of long-term memory including memory for facts and general knowledge, memory for things that happen in our lives, and skill memory that encompasses such things as knowing how to eat and walk. The form of memory that Ebbinghaus was interested in falls within the general knowledge category. He learned material within a particular content domain and then tested himself on what he learned. The key to Ebbinghaus' technique was his choice of learning material, what he learned so that memory content could be quantitated. This was a critical choice, and it represents a principal fork in the history of memory research.

One of the obvious attributes of long-term memory for general knowledge is that understanding plays a large role in what is remembered. There are numerous psychological studies that demonstrate this, but they mostly recapitulate what people experience daily. If you do not play chess, you can watch a chess game and not remember any of the moves that were made. If you do not speak French, you will not remember an overheard conversation in French. If you are like many people in the United States, you will remember little of hockey games except for the fighting. And so on. What chess, French conversation, and hockey have in common is that they are all meaningful. As understanding is the gateway to memory for meaningful experience, people who understand chess, French, or hockey will remember, often with exquisite detail, experiences in these domains. This circumstance makes memory for meaningful experiences difficult or impossible to quantitate. There is no reliable way to quantify understanding, even though every time a teacher assigns a grade, they are attempting to do just that. And consequently, there is no reliable way to quantify what people remember about things that are meaningful. This presents a dismal prospect to a nineteenth-century mind that is bent on bringing systematicity to the study of memory.

Insofar as meanings and understandings are resistant to quantitation, a quantitative law of memory retention necessarily requires content that is not meaningful. In what follows, material that is not meaningful will be referred to as nonsense – without the negative valuation that the term evokes. With this understanding, a quantitative law of retention will necessarily be a law for the retention of nonsense. This circumstance led Ebbinghaus to a consideration of what forms of nonsense might be useful in a memory study. Although there is now an enormous literature on the learning and forgetting of nonsense, Ebbinghaus was a pioneer and so had to come up with his own nonsense product. Initially he considered fragments from poems but rejected poetry on the basis that variation in the meaningfulness of the fragments would infuse variability into the measurement scheme. Eventually Ebbinghaus came to creating nonsense syllables, meaningless consonant–vowel–consonant clusters. Nonsense syllables solve many measurement problems all at once. They have no hold on the imagination, they have no place in language and so invite no linguistic complications such as word frequency, and they generate no sense of familiarity. Their utter meaninglessness promotes the likelihood that they will lead to reliable and reproducible forgetting rates. Second, each nonsense syllable is a discrete, isolated unit. A unit that is discrete and isolated has the property that it may be counted. Countability is the first step in quantitation, and it was a necessary step in the development of mathematical descriptions of forgetting.

The Concept of Working Memory

From Ebbinghaus' work onwards there have been two distinct strands of memory research. One strand has been concerned with memory for meaningful events and material. This strand has the virtue that it studies mind using materials that resemble the content that minds seem to be designed for. The other strand has been concerned with memory for bits of nonsense. This strand has the virtue that it permits quantitation and, frankly, it looks a lot more like science. On this second strand is found most of the psychological research into long-term memory and virtually the entirety of research into short-term or working memory. As nonsense has been foundational to the study of memory, a tutorial on nonsense will not be out of place in this discussion.

The Art of Making Nonsense

Experimental technique on the quantitative strand of memory research has long surpassed the nonsense syllable as a quaint and historically dated relic. Since Ebbinghaus, various ways of constructing meaningless experiences have been invented. These techniques all capitalize upon the property that meaningful experiences have narrative content. A definition of narrative content will not be attempted here, but narrative content may nevertheless be recognized by its having gist, an organizing principle or idea that underlies the specific words that are used to communicate the content in the first place. In essence, narrative content is compressible and that is what gist achieves, a compressed and concise retelling. In consequence, the art of creating nonsense comes down to manufacturing lists that cannot be compressed by an organizing principle or idea. There are two ways of making incompressible lists, by creating list items that are perfect nonsense in themselves, or by creating lists of meaningful things that have no relation to each other. An example of the first way is the paired associate. In paired associate learning, an item to be learned might be the conjunction of a number and a word such as in "dog-5." Dogs are meaningful and so too are numbers, but the conjunction is meaningless. In this way, a paired associate is very much like a nonsense syllable. An example of the second way is a list of words that are meaningful in themselves but appear random when placed in a collection. As the making of nonsense lists is something of a niche occupation, an example might be helpful. The following list appeared in a seminal article on working memory and was constructed to be maximally random (Lovatt, Avons, & Masterson, 2000): oblong, lemon, puppet, wizard, camel, tablet, coffin, and hammock. If a nonsense list of words does its job, the lack of association or relatability will achieve the same isolation and unitization that Ebbinghaus achieved with the nonsense syllable.

The Origin Story of Short-Term Working Memory

One of the ways in which the study of long-term and short-term memory are distinguished is in their respective reliance on the learning and subsequent forgetting of nonsense. The study of long-term memory is predicated upon a choice; study memory as it is used in meaningful contexts as a form of descriptive ethology or construct laws of forgetting that are based on the counting of retained nonsense items. The study of short-term memory, in contrast, is not offered this choice. Short-term memory phenomena, by virtue of their being short term, arise in circumstances where forgetting is almost but not quite immediate. Things that are quickly forgotten are generally without meaning, else they would not be quickly forgotten. Phenomenally there is a brief window of time in which almost any piece of nonsense is available for recall before it is forgotten. That brief window of time is created by something, and that something has been called short-term memory.

Prior to 1970, the concept of a short-term memory lacked definition; it was not at all clear what it was or how it worked. What was known about short-term memory was the product of isolated experiments employing methodologies that were uncalibrated and informal. An illustrative example is a rather famous study that investigated the role of rehearsal in the maintenance of nonsense lists (Peterson & Peterson, 1959). This study, occurring in the very early days, required a methodology that of necessity was made up from scratch. What the Petersons came up with was a method where each trial consisted of three consonants that were presented along with a number. Participants counted backwards from that number by threes or fours for a given period (between 3 and 18 s). The point of having people count backwards was to prevent rehearsal. Insofar as three consonants that do not form a word offer a meaningless experience, this methodology is guaranteed to produce rapid but not immediate forgetting – the exact conditions for studying short-term memory. In fact, participants forgot 80 percent of the consonants over a period of about 10 s, evidence that people will rapidly forget meaningless content unless allowed to rehearse it and or mentally refresh it in some way. That Peterson and Peterson were able to produce a forgetting function for short-term memory is testimony both to the power of experimental design and to the counting properties of discrete nonsense units. However, equally evident is that the empirical outcomes from this study are virtually uninterpretable. Because the measured amounts of forgetting were idiosyncratic products of the specific design choices that were made, the only thing about which we can be sure is that people will rapidly forget meaningless content.

The Serial Position Curve

An important development in the concept of a short-term memory was the production of the serial position curve in experimental contexts that seemed to be relevant to short-term memory function. A representative curve is illustrated in Figure 4.1. This curve resembles a shallow parabola and is intended to depict the quality of retention for a set of nonsense items that are presented sequentially. In sets of unrelated nonsense, the first few items and the last few items are better recalled than items in the middle; the so-called *primacy* and *recency* effects.

Without context, this curve may seem abstract and esoteric, but it is a common feature of the autobiographical memory for everyday events. Anytime a person experiences a series of events, meals at a restaurant, parking places at work, shopping trips to a market, that set might generate a serial position curve in long-term memory. It depends on whether the series is thematic, whether it has narrative content. If parking at work

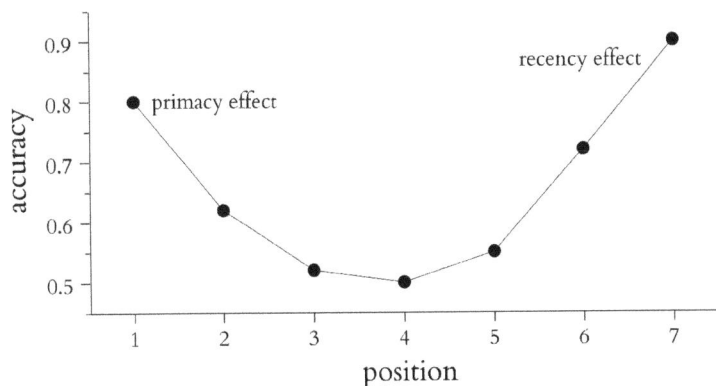

Figure 4.1

involves the randomness of finding the first open space, then there may be no relationship between successive parking spaces. When there is no relationship between successive events, the series itself is nonsense, just like a set of unrelated words is nonsense. When a set of successive experiences lacks narrative content, then the set satisfies the conditions for producing a serial position curve. For such sets, the first instance is likely to be memorable simply because the first time for any experience carries both novelty and distinctiveness. Last times may be memorable for reasons presumably related to their being last, as there is no subsequent experience that might create confusion. If an event in the middle is memorable, that is probably because the event was unusual – your meal was dropped, your car was broken into, you sprained your ankle, and so on. These examples make clear that serial position effects are a staple of personal autobiography and are in no way exclusive to the immediate forgetting that defines short-term memory.

Where serial position effects become relevant to short-term memory is when the set of nonsense experiences is compressed into a few seconds rather than playing out in the weeks, months, and years that comprise ordinary life. In a serial position experiment some small number of nonsense items, typically around five to seven, are presented. The item presentation is typically followed by immediate recall. The entire experience might consume maybe 10 s. The presentation and recall episodes would constitute a single trial and over many such trials a serial position curve in recall accuracy might emerge. The experimental 10 s curve and the year-long parking curve might both manifest recency effects, but in the case of the experimental curve, the recency effect has a special significance. The uptick in accuracy for the last few items is interpreted as implicating the existence of a memory buffer. The idea here is that this buffer is inherently limited in some way and can hold what is dropped into it for just a few seconds. What is in the buffer when recall is initiated are whatever pieces of nonsense were presented at the end of the list, the most recent items. As the items presented in the middle of the list may no longer reside in the buffer, perhaps they have evaporated or decayed, their recall is less likely. Finally, the items presented at the beginning of the list are the most remote in time when recall begins and so are probably not in the buffer either. But they might benefit from a primacy effect, the enhanced recall that being first bestows. In this way the serial position curve begins to tell a story, a story about a buffer that is volatile and short lived. This story reached maturity in the mid 1980s with the work of Alan Baddeley.

Alan Baddeley and the Phonological Loop

The concept of a short-term memory became considerably more sophisticated with the work of Alan Baddeley (Baddeley & Hitch, 1974). Baddeley advanced the theory of short-term memory by specifying its parts, how the parts worked, and how the parts interacted. The model was constructed around the core principle that short-term memory does a particular kind of work, that it is a working memory. The job of working memory is to support and enable higher-order cognitive functions such as reasoning, categorization, planning, and so on. Although Baddeley's conception of working memory mostly endures, his greatest contributions were empirical. Baddeley developed the kind of interplay between experiment and theory that is the hallmark of productive science. This is especially true of his work on the phonological loop. The phonological loop has historically been the most studied part within Baddeley's model of working memory, and consequently quite a bit is known about the brief retention of unrelated verbal material. The phonological loop is also likely what Fraisse is referring to when he remarks that short-term memory is not the capacity of apprehension.

Baddeley's phonological loop model was explicitly built to explain the recency effect and the maintenance properties of rehearsal. The model has as its foundation a phonological buffer into which learned material is dropped after it has been properly encoded. To explain the recency effect, the model proposed that immediate forgetting was due to a decay process that could empty the buffer in just a few seconds. The decay process was not itself explained, it was an assumption of the theory, but it does provide a specific mechanism for putting the "short" into a short-term memory. In this model the buffer is augmented by a process of rehearsal, the idea being that rehearsal reactivates the phonological encodings and sets them up for another round of decay. In this way the phonological loop creates a kind of mechanical system that resembles juggling; bits of phonology are always in the process of falling to the ground and will land on the ground unless plucked out of the air and tossed upwards for another round of falling.

The decay-rehearsal account of forgetting in the phonological loop had material impact on the field of working memory because it led to definite and testable predictions about memory function. One prediction that turned out to be decisive was the *word length effect* – that lists of long words will show greater forgetting than lists of short words. The prediction is entailed by the logic that as lists of long words must take longer to rehearse than lists of short words, long word lists will receive relatively fewer opportunities for maintenance and consequently more opportunities for decay. Baddeley's initial studies (Baddeley, Thomson, & Buchanan, 1975) did show the anticipated word length effect. However, it later became evident that words, although they might be stripped of narrative content in a list, are nevertheless meaningfully connected to language and culture. Words are actually very complicated objects from a memory perspective, much more complicated than other pieces of nonsense such as nonsense syllables, individual letters, and individual numbers. Memorability of words is influenced by number of syllables, frequency in spoken and written language, familiarity, imaginability, and so on. The importance of Baddeley's work inevitably attracted replications in which the words to be included in word lists received a withering focus (Lovatt, Avons, & Masterson, 2000). The upshot of this activity was that the word length effect failed to replicate when the word lists were carefully constructed so that word length was truly the only factor differentiating inclusion in the short and long list.

The falsification of Baddeley's theory of the phonological loop is a puzzling place to land. The basic phenomenon that working memory studies confront is a real one; people really do forget nonnarrative nonsense rapidly but not immediately. The problem is that there is no clear theory of why this happens. The theories that have been invoked to replace the phonological loop are so complex and so flexible in their construction that it is a lively question whether they are even meaningful. Perhaps the only conclusion to be drawn here is that the rapid forgetting of nonsense is very complicated. Thankfully, these are somebody else's problems. The path followed here leads to a very different conception of memory.

Memory With and Without Containment

The concept of memory that underlies everyday understanding is built on a metaphor, that memory is a kind of container. This metaphor feels so profoundly right that it may be difficult to conceptualize what memory could be if it did not involve the containment of memories. Part of the sense of rightness comes about from the experience of remembering, that recall is retrieval. Retrieval implies *retrieval from*, and *from* implies place, and that place is the container called memory. But part of the sense of containment also comes from our sense of the world, that we have heads, and inside of our heads is a brain, and brains are places that contain memories.

Beyond the metaphor of containment is the ontology of the things contained; the nature of what is in the memory container. This is potentially a difficult subject to contemplate, but in the context of working memory, it is quite simple. The concept of working memory has been built entirely out of recall performance for ensembles of unrelated *items*. What it means to be an *item* is to be discrete, unitary, and identifiable. Going back to Ebbinghaus' decision to use syllable clusters, discrete pieces of information have been foundational in memory assessment – this *number* of items were recalled, this *number* were not recalled. Without the ontology of particularity and item indivisibility there would be no counting, no numbers, and then there would be no serial position curve and effectively no literature on working memory.

These comments are not intended as a critique. They are intended only to be preface to a different and larger concept of memory. There are forms of memory that do not involve containers and are not designed to encode and store items of any type. As these forms of memory do not store anything, they also do not support remembering. Although such memory systems may seem foreign, they are in fact operating continuously in every moment of life, and they are precisely the kinds of memory that will be required to understand how groups form in time.

A more physical definition of memory systems will lead to ways of thinking about memory where storage and retrieval are replaced by the evolution of a dynamical system. In fact, the most general definition of memory does not mention containment or holding:

> A system displays memory whenever the current state of that system is influenced by past system states.

This conception of memory includes systems of containment and remembered content, but it also includes systems that embody memory without containment. Systems that embody memory without containment are generally physical, but such systems are also encountered in biology, ecology, economics, and virtually any domain where there is some kind of dynamics. Understanding memory without containment will make it clear how large the concept of memory is and will put the metaphor of containment into perspective. Eventually this discussion will lead to a conception of memory that will clarify how grouping works in time and why grouping creates a phase transition in time. To prepare for these developments it will help to have some examples of how container-less memory works in physical and biological systems.

Memory in Classical Mechanical Systems

Consider what is involved in throwing a ball. The motion of the ball is described by Newton's Law, $F = ma$ (force = mass × acceleration), but the law itself does not uniquely determine the trajectory. The trajectory is created not only by gravity, the F in Newton's law, but also by the initial angle and speed with which the ball is launched – what are referred to as the initial conditions. Where is the memory here? The memory is not in the ball, but in the trajectory the ball takes. The trajectory *embodies* memory without being a container by carrying the initial conditions forward in time. This is a legitimate and common example of how the present, the present location and speed, is affected by the past.

Another way of thinking about memory is through prediction; the trajectory of the ball can be predicted by knowing the initial conditions. Prediction and memory go together; only if past states influence future states can future states be predicted from past states. Classical mechanical systems are generally highly predictable because the initial conditions are preserved without any loss or distortion. Other systems, chaotic and stochastic systems particularly, display a spectrum of imperfect prediction. In these systems the initial conditions are lost over time, and it makes sense to speak of dynamical systems as having long- or short-term memory. Here long and short does not refer to different properties of containers, but to different dynamical histories.

Memory in Stochastic Systems

Any dynamic that incorporates probability will eventually lose track of its initial conditions. These systems are forgetful by nature, but how they forget and the rate at which they forget is an enormously rich area of inquiry in mathematics and physics. Here we will consider just the simplest examples to get a sense of how systems incorporate randomness while still retaining some degree of memory. Music is a particularly good vehicle for this discussion as it is a perfect example of a signal that displays an imperfect memory, blending surprise with anticipation.

Consider, then, two limiting ways in which random music might be generated. One limit is formed by creating what is referred to as a white noise. Here "noise" does not refer to a sound that is heard but refers rather to how a system that incorporates some degree of randomness behaves over time. A white noise is generated by processes that have no memory, where the state of the system in the present is independent of any state visited in the past. A roulette wheel is an example of a device that generates white noises to the extent that the wheel is fair, and past outcomes have no impact on the present outcome. There is the irony that virtually all people behave as if roulette outcomes can be predicted – after 10 black turns in a row is not red due? The number π could also serve as the source of a white noise. There are no detectable patterns in the digit sequence of π so there is no way in which any part of π could be predicted. An example of the time history of a white noise is shown in Figure 4.2. This top panel illustrates what the absence of memory in a physical process looks like. When the absence of memory is displayed as a contour it looks like a completely

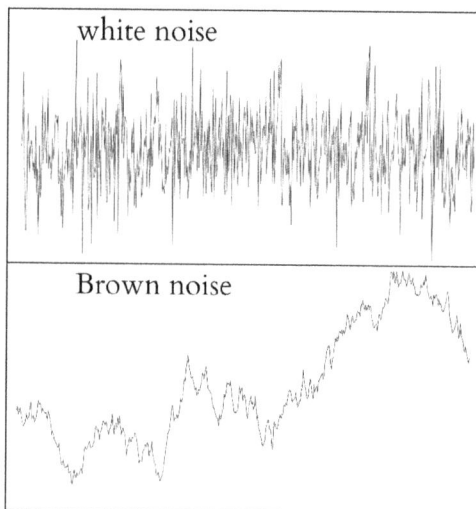

Figure 4.2

disorganized thicket. This contour rapidly oscillates without generating any trend whatsoever. It never strays far from where it started.

A white melody could be produced by literally playing π, assigning the digits 0–9 to notes. The corresponding tune would be unlistenable and almost surely annoying. Music that is constructed so that the present note is in no way influenced by past notes, where there is no memory, unfolds as a series of surprises. It will fail to create any sense of development and must also fail to generate feelings of tension and release. There is no question that much of modern music, both classical and jazz, became significantly whitened in the twentieth century. Free jazz and tone rows are intentional expressions of music without memory. These are interesting experiments but challenging for audiences.

The opposing limit to white noise is Brown noise, so called not to suggest a color but to recognize the physicist who characterized the microscopic motions of diffusing particles. Diffusing particles move along paths that are known as random walks, or drunkard's walks. This unflattering characterization of an inebriate's path imagines a person who takes a step to the right or to the left randomly. Diffusing particles are like drunkards in the sense that they are scattered to the left or right (or up or down) randomly. It might seem that if there is no preference for right or left that the drunkard or diffusing particle would never get anywhere. This is not true, the drunkard will ping-pong back and forth, but they will eventually diffuse to great distances, generating a path like that illustrated in the bottom panel of Figure 4.2. A contour representation of a Brownian noise has the appearance of ragged hills and valleys. Their raggedness comes from the ping-ponging that results when any step could be forwards or backwards. Nevertheless, the hills and valleys are the visual evidence that memory is embodied.

The sense in which a drunkard's walk embodies memory is that the drunkard's present position is highly influenced by past positions. Where the drunkard is at present is only one step away from where they just were. And that position is just one step away from where they were one step previous. And so on. So, if the walk is currently in a valley, it will stay in that valley for several more steps. And similarly, if the walk is on a hill, it will be found on that hill for some time in the future. This makes the drunkard's position highly predictable – for a while. Eventually a diffusion process will carry the drunkard or particle off to places that are not influenced by the distant past.

A musical example will give a good sense of what memory in a random process means. To make a random walk on the white keys on a piano, start on any white key on the piano, and flip a coin. If the coin lands heads the next key struck will be the one directly to the right, and if tails the one to the left will be struck. This process will create an exact representation of a drunkard's walk. Drunkard's walk melodies, by virtue of their construction, are guaranteed to sound uninteresting, not because they embody continuous surprise, but because they embody too little surprise. Every note is at most one step away from its predecessor. A Brown melody is too predictable, and this leads to the emotional experience of monotony. A Brown melody will eventually move across the keyboard, but only very slowly as it bounces back and forth.

Real music, the music that humans make and listen to, turns out, in a statistical sense, to be a blend of white and Brown noises. This blend is referred to variously as pink noise (deliberately misconstruing the person named Brown with a color), flicker noise, and $1/f$ noise. There is quite a bit that may be said about $1/f$ noises but here we shall just mention the discovery (Voss & Clarke, 1978) that if music and speech are treated abstractly as transmissions of pitch and loudness fluctuations, both streams are well described as $1/f$ noises. In their article they drew music and speech samples from four radio stations (playing classical, jazz and blues, rock, and news and talk). The details of how they characterized the time histories of the melodic and loudness content are wrapped up in spectral analysis, but their conclusion is not difficult to appreciate: the statistical character of music is literally a blending of white noise surprises with the slow drift of drunkard walk ambling.

Memory in Dissipative Systems

In the theory of dynamical systems there is a fundamental distinction between systems that conserve energy and systems that lose energy in unrecoverable ways. The distinction is important to the kind of dynamics that the system can display. Classical mechanical systems, in their idealized forms, are generally energy conserving: they neither consume nor shed energy. A consequence of energy conservation is the rather subtle attribute that they generally also preserve their initial conditions. Dynamically, every set of initial conditions launches a unique trajectory, and that trajectory remains distinct from all other trajectories launched from different initial conditions. In contrast, systems that do not conserve energy, dissipative systems, are organized by what are referred to as attracting states, or attractors. Attractors are states that capture trajectories regardless (mostly) of how those trajectories are initially launched. The idea here is that there might be a transient period while the trajectory settles down on to the attractor, but eventually all trajectories lose their initial conditions and express just the behavior of the attractor. When a trajectory converges onto an attractor it essentially has become amnesiac, where that trajectory came from is lost. The period of transience before convergence onto the attractor is essentially a dynamical form of short-term memory. This form of short-term memory will be useful in understanding phase transitions in grouping.

Attractors come in several forms. Fixed-point attractors are single states, and these are especially relevant to the dynamics of body organization. Homeostatic set points are, by definition, fixed-point attractors. A resting heart rate and a basal metabolism are states that the body invariably finds as it recovers from exertion. The most common fixed-point attractor in terrestrial physical systems is rest, or no motion. The state of rest is the attractor for any system that has friction and is not driven by an external force.

There are also a variety of cyclic attractors. Animal populations may, for example, converge to a cycle between two states: a high population state followed by die-off and a low population state followed by growth. Cycles which are attracting states are referred to as *limit cycles*; they are limiting states onto which a family of initial conditions converge. The feast/famine cycle becomes a limit cycle for animal populations when it is an inevitable outcome regardless of where the animal population starts off.

Finally, there are strange attractors. Discovered in the late twentieth century, strange attractors manage to be the solutions to deterministic physical equations – no probabilities involved – and still be unpredictable. The most common example of a strange attractor is the weather. The weather evolves according to Newton's laws and the equation of state relating pressure to density and temperature. Yet the weather cannot be predicted in the way that the trajectory of a rocket ship can regardless of how many weather stations provide data and regardless of the precision with which the current weather is known. Strange attractors create the butterfly effect where the slightest perturbations in the initial conditions eventually lead to highly divergent outcomes. The butterfly effect sets fundamental limits to how much can be known about future weather.

Perspective on Memory Containment and Gestalt

When Fraisse ponders how *bit* and *ter* sound, and when he concludes that whatever memory system allows them to sound like *bitter*, it is not short- or long-term memory, he is saying something important. Given the intellectual context that existed in his day, he is essentially saying that a memory system that brings things together, that builds the capacity of apprehension, is not based on containment. Containment is simply not a mechanism that can accomplish the being in relation that makes a group. Fraisse, however, does not offer an alternative form of memory for consideration, and this may reflect the limits of the context in which Fraisse was thinking and writing. The conception of memory that has dominated psychological theory is not designed to explain the perception of being in relation. Neither has psychology offered a viable theory of emergence – not in Fraisse's day and not now. Recognizing that the emergent property problem may never be solved, understanding the phase transition in time-based grouping will require a theory of memory that can separate the dynamics of grouping from the emergent property aspect of grouping. That theory will be developed in the following chapter through the introduction of dynamical forms of memory into Gestalt. The theory will only attempt to account for the proximity constraints that lead to the capacity of apprehension, the 2±1 s phase transition in group formation. The emergent properties of groups are a mystery that will remain.

CHAPTER 5

The Activated Mind

This chapter introduces a form of dynamical memory that can create the phase transition between being in relation in a group and being isolated in succession – Fraisse's capacity of apprehension. This form of memory is not static, it is not a container, but a time-dependent process. The kernel of this memory system is a decay dynamic and so this is a memory system that is inherently ephemeral. That it is ephemeral will ultimately provide the structure that is necessary to create a phase transition in time-based grouping. There is some irony in that the memory systems of interest here are already quite well known under the name of *priming* although there has been remarkably little effort in studying priming from the point of view of its being a time-dependent dynamic process. What little has been accomplished is highly relevant to understanding the phase transition in group formation, and consequently this chapter will begin with a good deal of experimental psychology. The experimental psychology is quite important and really the only thing that distinguishes this effort from speculative philosophy. The chapter concludes with some pictures that illustrate how decay processes naturally create phase transitions in both priming and grouping. These pictures are not sophisticated, but they manage to explain quite a bit of the phenomena.

Varieties of Behavioral Priming

There is a field of psychological research devoted to the study of memory phenomena that operate outside of awareness, that are not governed by the encoding–storage–retrieval paradigm, and which seem to have a distinctly physical character. Priming, as these forms of memory phenomena are known, has a very simple formal structure: past experiences change us in some way so that the present ability, tendency, or readiness to execute a behavior is modified. There is no remembering involved here. The priming event need only change us in some way so that going forward our abilities are somewhat different than they would have been in the absence of that earlier experience. This structure obviously has great generality and could include the ways in which childhood experiences influence adult tendencies and choices. The concept of priming is also clearly active when we say that someone is pushing our buttons. The mere existence of buttons implies that certain behaviors and feelings are activated by prior experience, and it is the case that buttons do tend to be far-reaching into the past. However, when cognitive psychologists study priming, their focus is on extremely subtle influences that are both well of outside of awareness and ephemeral.

The forms of priming that are studied in formal psychophysical experiments generally involve the subtle nudges that prior experience exerts on present behavior. These nudges tend to be so slight that they would not be evident were it not for the precision that can be brought to bear through the measurement of response time latencies. Typically shifts in readiness might involve only a few tens of milliseconds added to or subtracted from the time it takes to react and respond to very simple stimulus. This is evidently a very subtle form of memory, operating below conscious awareness. Nevertheless, every experience that a person has and every response that a person makes to the environment is wrapped up in priming. There is no more important form of memory as it is essentially what keeps us in close, very close, contact with the world and with our own bodies.

Activation is the predominant metaphor for the kind of memory that manifests as priming. The metaphor is intended to convey the idea that past stimuli and past responses may temporarily inject energy into the mind and bring it into an activated state. The idea here is that while the mind is in an activated or excited state, there will be small but detectable shifts in a person's readiness and ability to acquire certain forms of information and to execute certain forms of response. What the metaphor does not speak to is how activation works. The most important questions have to do with the dynamics of activation: what happens to activated states, how long do they stay activated, do different priming events have different activation dynamics . . . ? There are no psychological theories that are remotely able to address activation dynamics, but there are ways to answer these questions through carefully designed experiments.

The time course of activation is approached experimentally using a technique where behavioral samples are taken to produce snapshots of the activated state. The logic is illustrated in Figure 5.1, where a priming event produces a spike of activation which then decays. The dashed line depicts an idealized activation decay history that is intended to

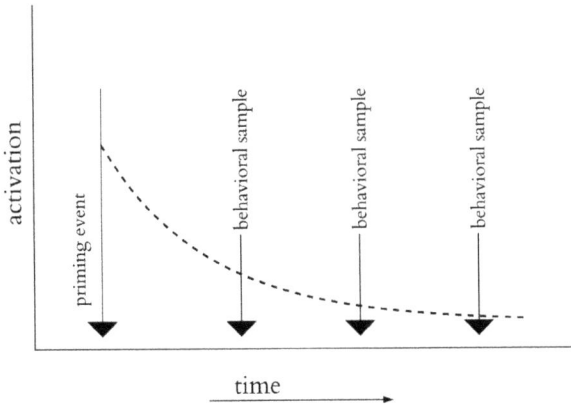

Figure 5.1

represent formally how the priming memory unfolds in time. The mental states depicted by the dashed line are not directly observable and must be recovered indirectly through the taking of behavioral samples. A behavioral sample will generally consist of some decision that the participant must enact as quickly and accurately as possible. The amount of activation that exists at any time following the initial priming event is inferred from the degree to which speed and accuracy have been changed from the baseline state – baseline being decision behavior that has no priming event in the recent past. In Figure 5.1 three of many possible sampling points are illustrated. Each of these samples would catch the decay process in a different state. In some ways, taking behavioral samples to measure activation is like measuring the temperature of a liquid using a thermometer. But there is a crucial difference. Taking a behavioral sample is not a passive observation. The behavioral sample is also a priming event, and it will generate its own activation history. So, once the sample is taken it is necessary to start over with a new priming event – or use the sample as a priming event. This fact guides how experimental designs are structured so that they can resolve the entire time history of activation decay.

Semantic Priming

Semantic priming is an excellent starting point for introducing not only how priming is studied but also the terms in which it has historically been

described and understood. Semantic priming is essentially the phenomenon that one thought leads to another, that thought is guided by association. The paradigm developed by Meyer and Schvaneveldt (1971) is
a seminal example of how an extremely subtle mental phenomenon such
as concept association may be given precise quantitation through an
appropriate experimental design. Methodology is critical in this context
as it is in every context where the subtle influences of priming are revealed.
Here the experimental wedge is provided by a lexical decision task. This
method involves the presentation of strings of letters, some forming words
while others form nonwords. The task for the participant is simply to
decide if the string is a word or not – as quickly and accurately as possible.
In a speeded choice task, the elapsed time between the presentation of
a stimulus, here a letter string, and the keypress response is known as
a response time latency. If everything works out as planned, patterns in the
latency data will reveal the state of semantic activation.

The method described here is a variant of the version developed by
Meyer and Schvaneveldt and is best illustrated by an example. A given trial
will begin with the presentation of a letter string which is always a word.
Let's suppose that this letter string is DOCTOR. Following this event
a second letter string is presented. Our focus is on the character of
this second letter string. There are three types of letter strings that might
succeed DOCTOR, illustrated here by BUTTER (unrelated word),
NURSE (associate), and BLICK (nonword). The issue here is how the
reading of DOCTOR influences peoples' abilities to solve the problem of
whether the second string is a word. What is invariably found is that the
time to figure out that BUTTER is a word or that BLICK is not a word is
not at all influenced by the previous reading of DOCTOR. However, the
time to figure out that NURSE is a word is substantially affected – that
time is much shorter than it would have been if the mind had not been
prepared by DOCTOR.

The interpretation given to this pattern of data is first that the concept
doctor creates some form of mental activation, and second that this activation moves along channels known as sematic networks. In this way of
thinking, *doctor* activation spreads to other concepts that are related to
doctors: *hospital, disease, surgery, medicine, nurse,* and so on. Because nurses
are conceptually related to doctors, the letter string NURSE is more
quickly recognized as a word following the reading of DOCTOR than it
would have been otherwise. Presumably the same benefits would accrue to
HOSPITAL and DISEASE. Now, because the concept *butter* is not an
associate of the concept *doctor*, we do not think about butter when

thinking about doctors, the letter string BUTTER does not get any of the spreading activation. BUTTER is solved as being a word precisely as fast as it would have been if DOCTOR has not just been read. The same logic applies to nonwords as nonwords are not concepts and so are not part of any semantic network.

Semantic priming provides a first opportunity for following the dynamics of an activated mind. The method is based on the sampling logic illustrated in Figure 5.1. First, the activated state is created by presenting a priming letter string, like DOCTOR. The activated state is then sampled by presenting a probe letter string such as NURSE or BUTTER after a variable delay. A concrete example might be the following chains of events:

presentation of DOCTOR – wait 1 second – presentation of NURSE.

In this example, the response time latency for solving the "is NURSE a word?" problem reflects the state of the semantic network one second after it has been activated. This two-step process would be repeated elsewhere in the experiment with different concepts (TIGER primes STRIPE but not CHAIR for example) at a variety of delays, say after 2 or 3 second waiting periods. Eventually this procedure will lead to resolving the dashed line in Figure 5.1, the full course of activation decay.

This is a relatively easy experiment to do, and it has served for many years as vehicle for teaching undergraduate psychology students about experimental design and data analysis. Figure 5.2 shows data from

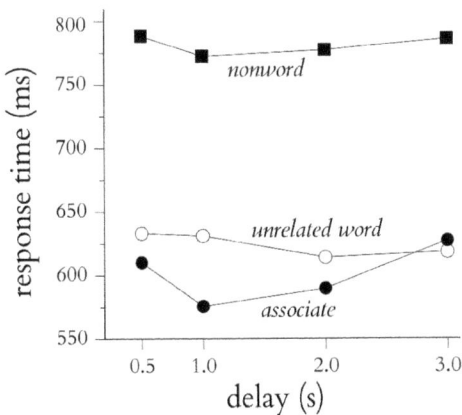

Figure 5.2

a typical experiment,[1] where response latencies to probe letter strings (such as NURSE, BUTTER, BLICK) are plotted relative to the delay following the priming letter string (such as DOCTOR). The plotted data points are the average response time latencies for deciding whether probe letter strings are words or not. These response times are divided into 12 different cells as there are four different delay periods and three types of probe string (associate, unrelated, and nonword). The story that Figure 5.2 tells is contained in the detailed shapes of the three curves. The top curve, labeled *nonword*, shows the average response time latencies for people to figure out that letter strings like BLICK are not words. The most important feature of this curve is that it is relatively flat, the latencies are independent of delay. The interpretation is that a concept like DOCTOR does not activate a nonword like BLICK. Consequently, it does not make any difference when BLICK is presented following DOCTOR; it does not participate in the dynamics of spreading activation and so the latencies are not influenced by delay. The same story applies to the middle curve labeled *unrelated word* that displays average response times to unrelated words, like BUTTER following DOCTOR. Because spreading activation is conceived as moving on semantic networks, a word like BUTTER does not receive activation from a word like DOCTOR. Unrelated words are like nonwords in this regard, and because they do not participate in the activation dynamics, it does not matter when they are encountered. Consequently, the length of the waiting period between prime and probe is immaterial, and this delay curve is also flat. The bottom curve labeled *associate* provides the key to understanding how the dynamic unfolds. There is important information present at every level of delay. At delays of a half second the effects of activation are already visible; associated words are recognized about 20 ms (milliseconds) faster than unrelated words. Activation reaches a maximal level at 1 s when associates are recognized about 55 ms faster than unrelated words. After a delay of 2 s, the activation appears to be waning, the difference between associates and unrelated words has diminished to 25 ms. And finally at 3 s there appears to be no activation; the *unrelated word* and *associate* curves have converged. In other words, after a waiting period of 3 s, a concept like DOCTOR no longer influences a concept like NURSE; NURSE has become the same as BUTTER as far as wordiness goes.

Figure 5.3 replots the data in a way that more clearly illustrates the time course of activation decay. The only aspect of the data that is relevant to activation decay is the difference in average response time latencies to associates and unrelated words:

Priming effect = unrelated word RT − associate RT.

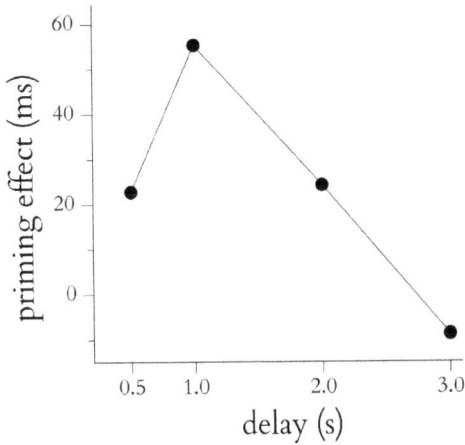

Figure 5.3

The greater the priming influence on lexical decision, the smaller the average response time (RT) for associates, and the larger this difference will be.

The curve displayed in Figure 5.3 tracks a concrete realization of the idea that the time course of activation can be recovered through delayed behavioral samples. Here there is an early surge in priming, peaking at a delay of 1 s, which is then followed by a decay epoch lasting only 2 s. That it is possible to peer into the structure of thought through psychophysical technique is not to be taken for granted, and it pays to step back and appreciate what this curve communicates. It tells us about the life history of ideas. Apparently, ideas decay after about 3 s. This measurement may provide a productive way of thinking about the rate at which the stream of consciousness flows.

Motor/Response Priming

We are now at the second stop in this tour of activation dynamics. At this stop the activation produced by a simple identification is followed through its decay history. The "what is it?" problem is not the simplest visual problem that might be given to a person. Simpler is the "did you see anything?" problem that is commonly solved in an ophthalmologist's office when retinal health is being assessed. But the simplest problem that can produce priming is arguably the "what is it?" problem.

Gilden (2001) measured the time course of activation in the "what is it?" problem by employing a methodology that was known to generate substantial amounts of priming. Circles, squares, and diamonds were randomly selected and presented with the understanding that the participant should identify them with a keypress as fast and accurately as possible. There is no simpler identification experiment: a shape is presented, it is identified by a keypress response, and then another shape is presented, and it is identified by a keypress response, and so on for over a thousand trials. The analysis of the activation state is, however, much more complex than that encountered in semantic priming. In semantic priming memory effects are confined to pairs of concepts: *doctor* primes *nurse*, *tiger* primes *stripe*, and so on. In this way the concept that is primed, say *nurse*, is activated by just one event, say *doctor*. Here the situation is more complex because there are no pairs. There is instead a sequence of shapes that appear on successive trials. The sequence of shapes generates a corresponding sequence of response time latencies as the participant churns through the experiment, responding *1* for circle, *2* for square, and *3* for diamond. What makes this paradigm so complicated is that any given response time might be influenced by many shape decisions made earlier in the sequence, all of them acting as priming events.

Responding to a shape is not as simple an affair as might be imagined. It involves four distinct stages: seeing that there is a shape (perception), deciding which shape it is (categorization), remembering which key has been assigned to the perceived shape (response mapping), and then pushing that key down (response execution). Each of these stages will undergo readiness fluctuations that depend in complicated ways on previous trials. To get a sense of how history might influence responding to a sequence of shapes, suppose that somewhere along the line, say starting at trial 31, the following three shapes occur as illustrated in Figure 5.4: circle on trial 31, square on trial 32, and then circle again on trial 33.

Here we wish to focus just on trial 33, the last member of this triplet. This trial exists in a particular context where the *circle* response (the 1 key)

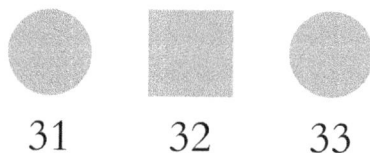

31   32   33

Figure 5.4

was activated, then the *square* response (the 2 key) was activated, and now it is necessary to activate the *circle* response again. It is the "again" nature of this history that is important. There is a general phenomenon in the allocation of attention that it is difficult to reallocate attention to a place that has just been vacated. This phenomenon is referred to as *inhibition of return*. In this context attention is being allocated not to places but to particular response mappings and responses. Yet the data are quite clear that it is difficult to return to a previously activated response if it has just been replaced by a different response. Response latencies to trials that have a history like trial 33 are always longer than average.

A second example drawn from another place in the trial sequence, say starting at trial 57, is illustrated in Figure 5.5: circle on trial 57, circle on trial 58, and circle yet again on trial 59.

In this case the circle on trial 59 is arrived at after two previous experiences of circles. At this point the response mapping for thinking "circle = 1-key" is warmed up, the finger that hits the 1-key is warmed up, and consequently the response time latency for identifying the circle on trial 59 will be shorter than average.

In reality, the ways in which the past influences the present in a sequence of shape decisions is much more complicated than these two examples might suggest. A shape decision that is made in the present moment might be influenced by, perhaps, the most recent five, six, seven, or more decisions. Just to be definite, suppose that shape decisions are influenced by the previous seven trials. If there are three shapes, such as in this experiment, then there are $3 \times 3 \times 3 \times 3 \times 3 \times 3 \times 3 = 2{,}187$ distinct histories leading up to the current decision. The myriad ways in which previous trials influence the present trial makes the characterization of the state of activation at any given moment impossibly difficult to describe. That, however, is not a flaw or much of a problem because we are not interested in an exact description of the state of activation so much as describing how the activation decays over time. The temporal dynamics may be studied using the same logic as used in semantic priming,

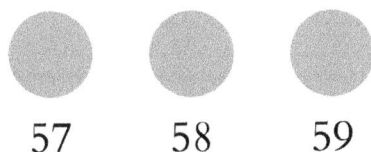

57   58   59

Figure 5.5

inserting time delays between events. In semantic priming the delay was placed between the priming word and the letter string that followed. In shape identification a delay is placed between the response to the previous shape and the presentation of the next shape. The logic here is that activation will decay during the delay periods, and this will lead to the response times becoming more uniform and less influenced by previous experience. Specifically, the inhibitory effects that cause trial 33 to be slow in Figure 5.4 will decay over time, and when trials are sufficiently separated in time, trial 33 will be responded to as if it had occurred in isolation. The same is true for the "warming up" effects in Figure 5.5. The activated state of being warm will decay, and when it does the last circle of the three in a row on trial 59 will be responded to as if it had occurred in isolation.

The complete picture of how motor/perceptual priming decays over time is illustrated in Figure 5.6. Here priming strength is plotted as a function of the delay between responses and subsequent image presentations – the response-to-stimulus interval. Priming strength is a statistic that was contrived (Gilden, 2001) to capture the effectiveness with which patterns like those illustrated in Figures 5.4 and 5.5 alter the speed of response. Unlike semantic priming, which required about a second to achieve full amplitude, it appears that motor/perceptual priming is maximal when there is no delay. Overall, the shape of this decay curve is highly reminiscent of radioactive decay. Motor/perceptual priming could reasonably be described as a decay process with a half-life of a couple of seconds.

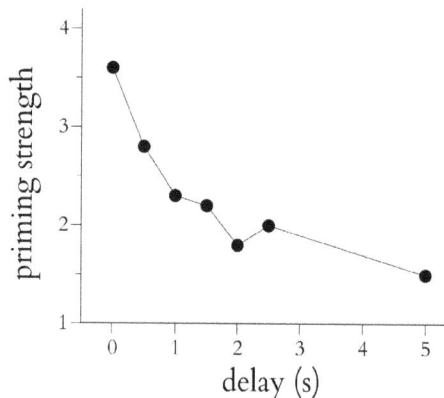

Figure 5.6 Copyright © 2001, American Psychological Association

Negative Priming

The third stop on this activation tour is negative priming. This is an interesting form of priming insofar as it is inhibitory, reducing readiness for response, and the paradigms that have been invented to study it look for increased response time latencies – the opposite to how the lexical decision task works in semantic priming. Negative priming arises from the yin/yang logic of selection; selecting *this* means not selecting *that*. Selection and deselection operate together, and a moment's thought will make clear that almost any mental act that involves attention or choice will have this quality. Attention and choice generally imply focus and focus necessarily implies exclusion. What the negative priming literature teaches is that those things not chosen or not favored by attention are not out of mind. They are very much in mind, but in a suppressed state. In the context of negative priming, measuring the time course of activation comes down to measuring the time course of the release from suppression.

There must be an infinite number of ways in which suppression might be investigated. The ways that end up being enacted and reported in the literature are invariably idiosyncratic and so always a little odd. The study of interest here is no exception. It was chosen for this discussion not because it is particularly insightful or illuminating, but because it measured the time course of suppression release, and this is a very rare occurrence in the published literature.

Neill and Westbury (1987) measured a form of suppression that naturally occurs in the Stroop effect. To appreciate what they did it is first necessary to understand what the Stroop effect is, and then how the experience of the Stroop effect could generate a suppression of readiness that could be carried into the future. The Stroop effect takes many forms, but the original form, and that used in the present study, involves identifying colors, solving the "what color is it?" problem. Reporting on a color turns out to be a little fraught if the object that is colored is itself a color word. Obviously, this is likely a situation that will only arise in a psychology experiment. Nevertheless, Figure 5.7 shows examples of easy and difficult "what color is it?" problems; the word *black* written in black and the word *white* also written in black. The Stroop effect is simply that it is easier and faster to name a black color when it is applied to the word *black* than when it is applied to the word *white*. In the first case the color and word are congruent whereas in the second they are incongruent. The Stroop effect is so robust that it does not require a carefully controlled

black   white

Figure 5.7

laboratory experiment for its demonstration. All that is required is a list of, say, 20 incongruent color/word combinations and a second list of congruent color/word combinations. The congruent list will be read fluently in seconds, whereas reading the incongruent list will be a real struggle.

There are scores of articles on the Stroop effect and so there is more than one explanation of what causes it. Nevertheless, the simplest explanation is that people are highly practiced at reading but rarely practice color naming. With as much practice as people have at reading, the word comes to mind instantly and automatically. Naming colors, in contrast, is neither instant nor automatic and is often quite difficult. This circumstance gives the word that is read a head start toward a response even though the word itself is irrelevant to response; the task is to name colors, not words. Now consider what this head start means for response. In this example there would be one key assigned to the black color and a different key assigned to the white color. For the word *white* written in black color, the keypress for white color must be suppressed even while it is being prepared. In essence, the participant goes through a "not white" experience so that they can execute the correct black color response. Suppression of the read word and the substitution of the correct color response takes time, several tens of milliseconds. Those tens of milliseconds end up lengthening the time it takes to solve the color naming problem when the word and color disagree. In contrast, for the word *black* written in black color, there is congruence between both color naming and reading. In this case there is no moment of "not black" suppression and no time lost.

The Stroop effect involves suppression, but it is not regarded as negative priming. In the interpretation given here, the read word does not suppress or inhibit the ability to color name, it is just that the read word is a racehorse and gets a response to the fingers before the color is able to mount its own response. Negative priming comes into the Stroop task when a response on one trial negatively affects the ability to respond on a subsequent trial. The situation that Neill and Westbury were interested in was when the irrelevant color word on one trial is the relevant color on the next. Unquestionably the logic of how suppression is carried forward is confusing, necessitating a concrete example.

Consider a design that includes three colors (blue, green, red) and three words (*blue, green, red*). Each of the three words can be painted in three different colors so there are nine different Stroop stimuli in all. In a potential experiment, colored words are randomly selected from this pool of nine, say, 100 times. The participant's task is just the Stroop task – respond to colors, and this they do 100 times. Focus, now, on one of the words, maybe the 34th. That word turns out to be the word *blue* written in red. The participant is trying to name colors so this is an example where suppression of the "blue" response must occur so that there is time for the correct "red" response to make it onto the keyboard. It is here where negative priming may come into play. Let's say on the 35th trial the word *green* colored in blue is encountered. This is another example of the Stroop effect, except now "blue" is the correct response because the color is blue. But the "blue" response was just suppressed and so is now less available than it would have been otherwise. The "blue" response was already destined to be slow because of the Stroop effect, but now it is slowed further by lingering suppression from the previous trial. Lingering suppression is a form of activation, and it will decay. Neill and Westbury recognized this, and they measured the time course of decay using the same delay insertion method that was used in motor/response priming. The release of suppression with increasing delay provides an effective way of visualizing the time course of suppression decay.

The decay curve that Neill and Westbury measured is illustrated in Figure 5.8. The *y*-axis plots the suppression effect while the *x*-axis plots the time delays (0.02, 0.52, 1.02, and 2.02 s) that were used to obtain snapshots of the decay process. The suppression effect is intended to measure how much color naming is slowed by lingering suppression, considering the slowing that would exist anyway due to the Stroop effect. Like the semantic priming effect, the suppression effect is a response time difference:

> suppression effect = time to color name when there is lingering suppression –
> time to color name when there is no lingering suppression.

Because suppression lengthens response time latencies, this way of formulating the suppression effect guarantees that it will generally be positive.

The time history of suppression has much in common with the time history of semantic activation. Suppression peaks early, here reaching a maximum at about 0.5 s, and then enters a period of decay. There are

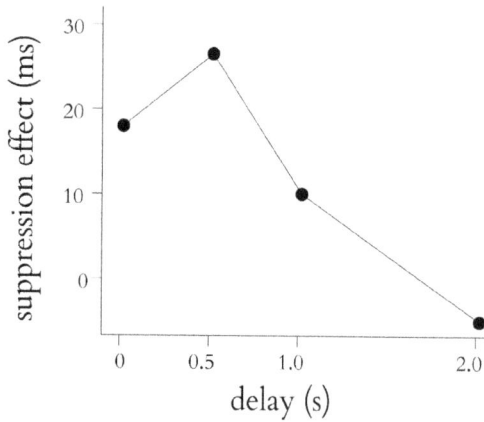

Figure 5.8

also differences, the most important perhaps being the relative weakness of negative priming in this Stroop paradigm. The suppression effect never exceeds 25 ms, about half of the effect found in semantic priming. That Stroop negative priming also decays to zero more rapidly than semantic priming, about a second sooner, may be related to its relatively low amplitude.

Perspective on Priming Decay

Both time-based grouping and priming appear to have a phase transition near 2±1 s. In both cases there is a relationship of some kind that is established when neighboring events are in close proximity, and a form of succession when they are not. The way the relationships manifest are, however, quite different, and this has profound implications for what is known about their dynamics. When two or more things come into a grouping relationship there is always an emergent property that enters experience. This property will generally be quite vivid and present in awareness. It is the self-verifying nature of vivid experience that allows Fraisse to simply present *bitter* and *tick-tock* to his readers in making the point that time-based groups fall apart when separated by just a couple of seconds. In essence, Fraisse does not present an argument, he simply reminds us of what we already know. The situation could not be more different with priming phenomena.

The relationships that form in priming do not lead to emergent properties. Rather, they just lead to temporary activations or suppressions. Priming makes us a little more or a little less ready to do something and the key modifier here is "little." Priming effects are so subtle that they require an experimental design and a measurement tool like response time to tease them out. The most profound priming effects, such as the speed-up in lexical decision given to NURSE by DOCTOR, is still only a few tens of milliseconds and is well below the threshold of awareness. It is simply a fact of mental life that spreading activation is not vivid. Also not vivid is the lingering suppression in Stroop, nor the finger response activations in motor priming. Nothing would be known about these forms of priming were it not for the development of psychophysical technique.

The fact that priming effects are not vivid but must be teased out has implications at the level of scientific culture. The first is that measurements of priming lifetimes will be rare. A priming lifetime will be measured when somebody sets out specifically to do so – it must be a conscious effort. Just thinking about DOCTOR and NURSE will not lead to any conclusions about the lifetime of semantic priming. It is necessary to create hundreds of semantically related pairs, hundreds of semantically unrelated pairs and hundreds of nonwords. Then it is necessary to place all these words into an experiment manager, run dozens of participants, and then analyze a raft of data. All of this requires a specific intention. Over the history of experimental psychology, the intention to measure a priming lifetime is rarely encountered. The three examples of priming lifetime presented here are not representative of a class, they are the class.

The second implication is that priming lifetimes, because they involve an experimental methodology to discover, will necessarily be bound to the many decisions that were made in its development and fabrication. It is unavoidable that when a psychological effect is brought into view by an experimental methodology, what exactly comes into view will depend critically on the methodology. This is especially relevant for the measurement of priming lifetimes because the phenomenon of priming is not experienced in a way that it may be reflected upon – everything is happening at a subconscious level. In situations like this, it is not uncommon for different investigators to make different methodological choices and to arrive at different conclusions. Where there is no fundamental theory and the phenomena being studied are outside of awareness, there will be controversy.[2] A consequence of this situation is that what is known about priming lifetimes will be sparsely supported and methodologically bound. It is inevitable that what is known about priming lifetimes will be weak compared to what is known about grouping lifetimes.

Theory of Priming Decay

The evidence from these three priming paradigms is that whatever it is that makes priming happen, priming decays over the course of a few seconds. The diversity of the paradigms, the methods employed, and the kind of data collected make clear that the decay behavior is generic and does not require special circumstances or preparations. We are going to take these results as given and create a general framework for conceptualizing the dynamics of priming – and just the dynamics. In other words, there will be no attempt to understand anything about priming except that it occurs and that it has a particular way of unfolding. The separation of the dynamics of priming from the specific manifestations of priming is what makes a study of temporality possible. The same separation will prove to be essential in theorizing about the temporal dynamics of grouping.

The framework developed here is based on the single idea that events create mental activation. What counts as an event is intended to be very general. Almost anything that happens can serve as a trigger that will eventually influence future thoughts, percepts, and acts. The concept of activation is intended here to be something physical, a form of energy or juice or heat. What activation is not, what it is being distinguished from, is information. So, although there is information to be derived from events, the activation produced by contact with an event is to be thought of as something that is devoid of meaning – ecological meaning or any other kind of meaning. In this way, the mind is conceptualized as a medium that is brought into activated states through intermittent injections of energy by world events, and priming is the phenomenon where an activated state modifies an organism's capacity for thought and behavior.

It is here that activation dynamics become relevant. In physical terms, activated states shed energy and cascade through a series of intermediate states until they reach a ground state. In the simplest version of this framework, both the energy cascade and the ground state are generic. This does not mean that priming is generic – it manifestly is not, as there are many forms of priming. It means only that the way activation is shed is generic. In this conception, there is just one way for a mind to be at rest, one ground state, and regardless of how a mind is activated, it will return to the same state of rest. From this point of view, priming is a transient state of mind, a brief waystation on the way back toward the resting state.

In physics the state of rest would be referred to as a fixed-point attractor. This may be a useful way to understand the decay dynamic. The idea here is quite simple; the mind wishes to be at rest, but the world continually buffets it with activation. Despite this, the mind continually sheds that activation and is continually returning to its preferred state. Now, if all transient activity is reducible to a single process of decay, then the dynamics will be governed by a single characteristic time, a decay timescale. This seems to be observed insofar as semantic priming, motor/response priming, and negative priming all seem to be decaying on the timescale of 2±1 s.

An illustration of the theory is given in Figure 5.9. Starting from the left and moving to the right we encounter a mind that is in its resting ground state. At some point the mind receives some activation. Here the activation is created by the letter string DOCTOR. Activation, being conceptualized purely in physical terms, is transient and immediately begins to decay. During the period of decay there are opportunities for the transient activation to make itself known. If the letter string NURSE is presented at some point during the period of decay, the mental processes that turn letter strings into recognized words will be influenced by the state of activation. The influence will be large if NURSE arrives early in the decay process and will be less if presented later in the decay process. These influences may be measured, and the graph of influences at early and late appearances of test strings like NURSE will lead to a decay curve for semantic priming. Exactly the same picture could be made for negative priming or for motor/perceptual priming. Decay curves in data are simply mirrors of an activated state transitioning to a ground state.

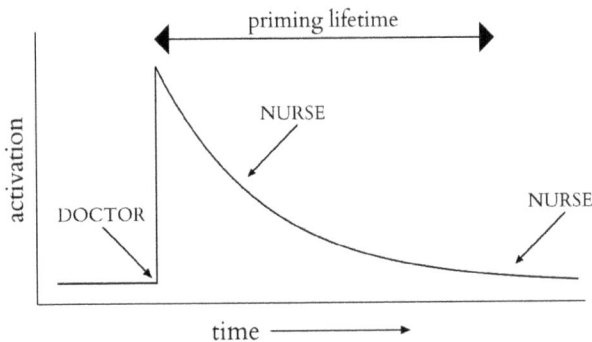

Figure 5.9

Theory of the Phase Transition in Group Formation

The theory outlined here relies heavily on the observation that grouping in time and priming have similar temporal structure even though they are radically distinguished in how they are experienced. First, they both operate using a dynamical form of memory where the past influences the present, not by storing and retrieving, but with the building of bridges. Bridges of the priming type create activations and inhibitions. Bridges of the grouping type create relations and "otherness." Second, and most importantly, both types of bridge have a brief maximum span. The experiments presented here converge on the empirical finding that bridges of the priming type have maximum spans of a few seconds, whereas everyday life supplies ample evidence that bridges of the grouping type span about 2±1 s. These commonalities suggest how a theoretical foundation for time-based grouping might be built.

To begin we will conceptualize the phase transition of 2±1 s as reflecting the lifetime of a transient pulse of activation. The idea is illustrated in Figure 5.10 through reference to Fraisse's wondering how *bit* may be followed by *ter* and still sound like *bitter* so long as the second syllable is not delayed by more than a couple of seconds. Figure 5.10 has the same elements as Figure 5.9, which depicts the activation structure of priming. The resemblance is, of course, intentional as the return of transient activation to the resting state is conceptualized as being a global mental process. In the Gestalt context, instead of DOCTOR creating a pulse of activation, the pulse is created by a syllable, *bit*. Eventually the second syllable, *ter*

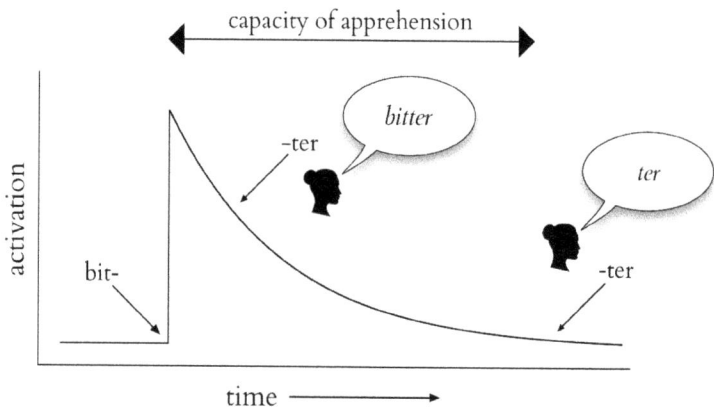

Figure 5.10

arrives. If it arrives before 2±1 s elapses, it will be grouped with *bit* to make *bitter*; *bitter* appears in the thought bubble. If *ter* arrives after 2±1 s has elapsed, it fails to bind to *bit* and arrives as just another moment in a succession of moments. Succession is depicted by the relegation of *ter* to its own thought bubble. In this way *ter* operates like the letter string NURSE, sampling whatever activation remains from the decay process. But instead of the lexical access of a letter string being influenced by residual activation, the residual activation influences whether *ter* is bound to *bit* or whether it floats free.

There is a deeper way to conceptualize group formation. The picture so far developed is not complete because it appears as if *bit* and *ter* are playing separate roles, as if *bit* is a pitcher and *ter* is a catcher. A more general picture treats *bit* and *ter* as simply two generators of a dynamic where activation decays toward a fixed-point attracting state of rest. In this case their coalescence into *bitter* is conceptualized in terms of activation overlap. Figure 5.11 illustrates the concept. The filled circle depicts where the two activation histories first intersect and where activation overlap begins. In this way of looking at grouping, a necessary condition for group formation is simply that there be sufficient activation history overlap.

The concept of activation overlap provides a general temporal template for understanding time-based grouping. In Figure 5.12 the template is applied to the formation of a melody. Here the first four notes of a traditional spiritual are noted on the staff together with the lyrics. Associated with each note is a transient activation decay curve. These notes are imagined arriving with sufficient alacrity so that every note overlaps with at least its closest neighbor. The conditions for grouping

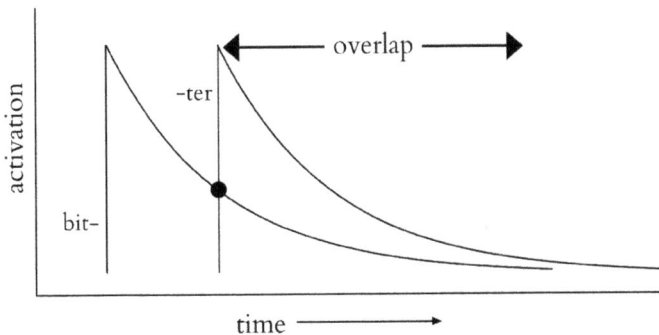

Figure 5.11

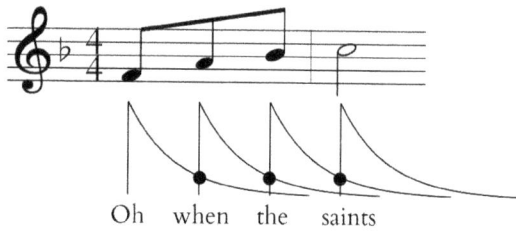

Figure 5.12

are met and these notes will make an emergent ascending contour that sounds like something associated with Louis Armstrong. The template of activation overlap is simple, but it must be. The ubiquity of the 2±1 s phase transition implies a common mechanism, one that does not incorporate any property of grouping beyond arrival times. This is what the template achieves.

Progress through Separation

To the extent that any progress has been made in this inquiry, it is attributable to the narrowness of its focus. There has been no attempt to understand anything about priming or grouping beyond how they operate over time. This strategy is justified if the process of activation decay operates independently of the processes that create groups in Gestalt, and which modify behavior in priming. Independence assumes that grouping and priming processes take whatever activation is given to them and then use that activation to do Gestalt and priming things.

Separating the dynamics of priming and grouping from what priming and grouping accomplish is essential to this project. There may be no phenomenon more global or more complex than that found in the temporal bridges built in priming and grouping. The forms of priming are virtually infinite, as anything that touches the mind or body can potentially create an activated state. Also infinite are the forms of emergence and all of them defy the kind of description that passes for analysis. If there is any insight in this project, it is recognizing what is intractable and what offers opportunity. The dynamics are tractable, especially as conceived here, where activation decay is regarded as simply a return to an attracting state of emptiness and rest. This strategy will prove its merit by the experimental psychology that it leads to.

Notes

1. There is also unpublished data on the time course of semantic priming that appeared in a conference proceeding (Meyer, Schvaneveldt, & Ruddy, 1972). It is noteworthy that the people who first described semantic priming were also the first to measure its dynamic and yet did not regard the dynamic to be of sufficient importance to publish their decay curve.
2. Hasher et al. (1996) examined the time course of negative priming in a spatial paradigm and found no evidence for any diminution in the suppression effect over delays as long as 2.5 s. It appears, however, that the experimenters placed pieces of tape onto the computer screen to mark potentially relevant positions – an innocent but potentially decisive decision. The pieces of tape must direct visual attention, and this alone can explain why there was a failure to find decay of negative priming.

CHAPTER 6

The Decay Process

The activation template that leads to temporal constraints on grouping and priming is built around the notion of decay. The centrality and importance of this construct imply that the template is only as well defined as is the decay concept. The idea of decay is not itself problematic, but so far it has been treated just in the informal sense of becoming less, diminishing. Activation, however, is intended here to refer to a physical process and there is an opportunity here to think about decay from a physical perspective. A physical perspective will, perhaps unsurprising, lead to a physical description. This description will bring definition and specificity to the activation template, but more interestingly it will send the inquiry into uncharted territory.

Decay generally refers to any process that takes a system from a state of higher energy to a state of lower energy. What it might mean specifically for transient mental activation comes down to specifying the path that the mind takes as it sheds transient activation and attempts to return to a state of rest. To the extent that the return to the attracting state is a lawful process, this path will have a definite shape, and that shape might be describable in mathematical terms. Getting to the point of definite shapes and mathematics will be challenging as there are no dynamical theories in psychology and certainly no equations that are known to describe the dynamics of activation. There is little recourse but to acknowledge this state of affairs and to inquire into how it might be improved.

In the most elementary terms, decay is a pathway that takes a system from one state (A) to another state (B). The picture at this point is just a generic downward trajectory such as that illustrated in Figure 6.1. Our goal here is to take the generic path and create one that is specific, and which has some theoretical motivation. To derive this better path, we are going to need some principle or strategy that constrains and limits the possibilities.

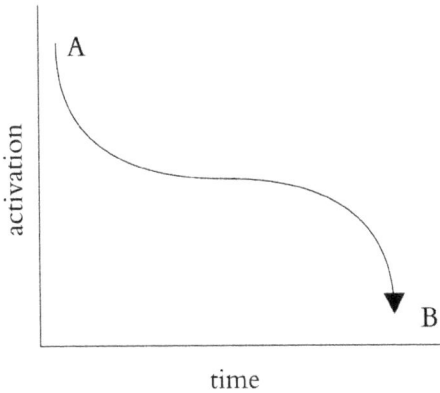

time

Figure 6.1

The Shape of Forgetting

Historically, forgetting over time has been thought to follow a power law. As forgetting is a type of loss, the power laws that pertain to loss have expressions such as

$$M(t) = M(0)^* t^{-b} \; b > 0,$$

where $M(t)$ is intended to convey how much memory content remains at any given time, t. $M(0)$ is the memory content that originally existed before forgetting commenced, and b is some positive constant that is the power in the power law. There are many other forms for generalizing the power law that do not blow up at time $t = 0$, but this equation has the virtue of being simple, and it is sufficient for the purpose of introducing what forgetting looks like. This expression, evaluated for $b = 0.5$ and $M(0) = 1$, has the graph illustrated in Figure 6.2. The appeal of the power law principally lies in the fact that forgetting curves, going back to Ebbinghaus' forgetting of nonsense syllables, often do look like Figure 6.2: steep memory loss early with slow memory loss later.

Still, even if human forgetting data does look like a power law, that does not mean that the forgetting follows a power law. Something more persuasive is needed to make a case for any law. If psychology were physics, the power law might be an entailment of a fundamental theory of brain or mind. However, as there are no fundamental theories of brain or mind, the power law is not something that might be deduced. The case for the power law being a true description of forgetting must be made in terms of how

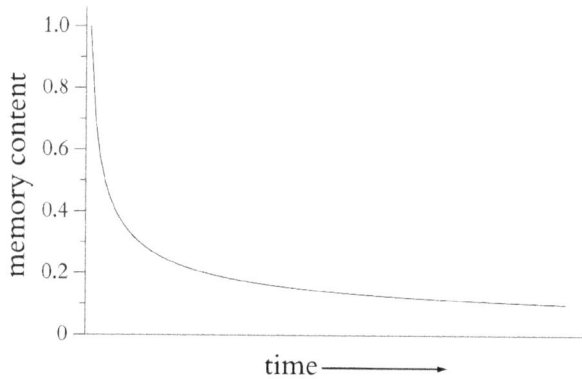

Figure 6.2

well it explains forgetting data. The enterprise of explaining psychological data invariably involves the construction of mathematical models.

Mathematical models are a principal tool in cognitive psychology, and some comment is required to understand what they are and how they work. Mathematical models are, fundamentally, frameworks for characterizing data. In this role they act to distill complex systems of data into much simpler structures: mathematical expressions.[1] The distillation is successful to the extent that the model and the data resemble each other, that they have the same shape. And yet, it is at this juncture that modeling becomes devilishly complicated. What does it mean for a model and data to share a common shape? In psychology this has come to mean that the model fits the data. So far so good, but what does it mean to fit data? This question is the point of departure into the seamy underside of how data are treated in psychology.

The most important feature of mathematical models in psychology is that they are incomplete and only find their full expression when they are fit to data. The power law, for example, has an exponent, b, which has an unknown value until the power law is fit to a data set. The idea here is that the power law has an infinity of possible shapes because there is an infinity of possible b values, whereas the forgetting data have only the one shape that was produced in a forgetting experiment. The goal in fitting a power law to forgetting data is to find the one exponent, the one b, that makes the power law most closely resemble the memory fall-off in the data. Because b is unspecified until it encounters data, it is referred to as a *free parameter*. Free parameters are one of the defining characteristics of psychological theory and their appearance is a signal of a science that lacks fundamental law.

Modeling using free parameters looks vaguely like testing physical theories using data, but it could not be more different. In the testing of a physical theory, the theory is stated and then it is compared to experimental data. If the theory and data do not agree, then there is either something wrong with the theory or there is something wrong with the experiment that collected the data. Modeling reverses the order of events. The data are collected, and then the model is fit to data by adjusting the free parameters. This means that models might not have a chance to disagree with data to the extent that the fits are decent. It is inherent in the mere existence of the free parameter that models will generally be able to shape-shift to resemble the trends in data. Because the data are used to make the model, the data cannot be used to reject the model in the way that a physical theory may be proved false. The entire issue of how data are used to test theories has, as a result, become quite fraught in cognitive psychology. The truth is that the power law of forgetting is not an established law. It is a model.

The Meaning of the Power Law

Mathematical models of psychological data acquire meaning by embodying principles that have psychological significance. The power law description of forgetting has been important not just because it fits data, but also because it has the entailment that older memories are more resistant to forgetting than more recently formed memories. This interesting and somewhat odd property is often referred to as Jost's 2nd law. That the power law is protective of old memories is not obvious from the way a power law looks, but it is there in the way the forgetting curve flattens out at long times. The flattening implies that following a period of substantial forgetting, whatever remains has staying power. The staying power of surviving memories is, in fact, the defining feature of a power law, but to see that will require a look at the rate equation for memory loss that has the power law as a solution. Now, it may be the case that books on human temporality generally manage to avoid rate equations, but here they are essential. It is not possible to really understand models of forgetting without understanding the factors that influence forgetting. It is only by examining the forgetting rate that these factors will be clearly laid out. Rate equations will make several appearances throughout this book as without them there is no path forwards. Ultimately it will be the close examination of rate equations that will settle the shape of activation decay.

The construction of a rate equation for forgetting begins as an idealization where memory is conceptualized as a kind of reservoir and forgetting as a kind of draining. The contents of this reservoir might be thought of as memory content or memory strength. A rate equation for forgetting might be written:

$$\Delta M/\Delta t = \text{Model},$$

where "$\Delta M/\Delta t$" should be read as "the rate at which memory content depletes," and "Model" here refers to whatever is conceptualized to influence the rate of forgetting.

The notation $\Delta M/\Delta t$ may be unfamiliar, but it is just a rate. The notation Δt refers to a time interval and ΔM is how much M changed during that interval. If M were distance, $\Delta M/\Delta t$ would be speed. As M is conceived to be an amount of memory, $\Delta M/\Delta t$ is the speed at which memory content is lost.

Different models will fill in the right-hand side of this equation in different ways. The simplest right-hand side that leads to the power law is:

$$\Delta M/\Delta t = -\,b^* M/t,$$

where b is a constant that ends up being the power in the power law. A minus sign has been put in because this is a forgetting process where $M(t)$ is getting smaller. The rate equation has two pieces, one piece is the M, and the other piece is the $1/t$. The $\Delta M \sim M$ part of the equation expresses the idea that the forgetting rate is proportional to the amount of memory content that is available to lose. This dependence is not remarkable, and it is in fact difficult to construct examples where it is not true. For example, over an hour which will evaporate more water – a lake or a teacup? This proportionality is not expressing a deep principle of memory so much as a principle of systems that are aggregations. In this sense a lake is an aggregation of water and human memory is an aggregation of information. The rate equation expresses the general rule that systems that are aggregations have loss rates that are proportional to the size of the aggregate.

The second component, the part of the rate equation where $\Delta M \sim 1/t$, is the interesting part. This term has the implication that the forgetting process behaves as if it knows how long it has been running. When t is small, forgetting is young and it runs strong just because $1/t$ is large. As time marches forward and the forgetting process ages, $1/t$ becomes progressively smaller as t becomes progressively larger. In this way the brakes are put on the forgetting process. To be clear, there are two ways in which the forgetting rate is throttled. Because the forgetting rate is proportional to the amount of stuff available to forget, as time marches on there will be less to forget and so the rate will slow. The $1/t$ term, however, generates a different and independent pathway for slowing the rate of forgetting. When time appears explicitly in the rate equation, old memories are treated differently than young memories. If a memory manages to survive forgetting it will, just by virtue of its older age, be less forgettable. Another way of saying this is that if two memories have the same strength, M, but have different ages, then the older one will decay more slowly than the younger one. This last sentence is an almost literal restatement of Jost's 2nd law that dates from 1897.

The issue here is not whether the power law is a true description of human forgetting or whether Jost's 2nd law is true. These are difficult questions about the nature of cognition and must be answered with data. What confronts us here is the much simpler question of whether the rate equation that leads to the power law should be regarded as a plausible description of activation decay. Answering this question comes down to deciding whether activation decay as it appears in the grouping and priming templates might have age dependence. If not, the path from A to B in Figure 6.1 cannot be a power law.

The conception of a memory that involves storage and retrieval also quite naturally involves the assignment of an age. Anything that might be dredged up in an episode of "I remember . . . " has an age dating from the event in the world that was memorialized. Activation, however, is not stored and it is not remembered; there is no "I remember . . . " for, say, when lexical activation is transferred to NURSE from DOCTOR. Although activation does follow a dynamic, there are no activation memories and consequently there are no activation memory ages. Activation is not a thing that can be young or old; as time passes, it does not age, it just becomes less. If activation does not have an age in the way that memories have an age, then the $1/t$ piece will not be part of its rate equation, and however activation decays, it will not decay as a power law.

Physical Decay Laws

As the goal here is to find a physically acceptable law for activation decay, it might be tempting to hold a model competition in a fitting contest on whatever decay data are available. Fortunately, or unfortunately, this will not be taking place. There is the practical problem that there are very few data sets that could be entered into the competition. The three examples of priming decay that have been discussed have illustrative value only. They make a point, that activation decays, but they would have little utility in a mathematical analysis of shape. Then there is the logical problem that activation is never directly observed, only the fruits of activation as they appear in human behavior through priming and grouping. This situation is very much like Plato's cave allegory (*The Republic, Book VII*) where we cannot know of the things that truly exist, only the shadows they cast. We can see the effects of activation decay, but we cannot gaze upon the decay process itself. It appears, then, that the problem of finding an acceptable decay function must be approached obliquely.

One starting point is to consider what decay might look like if we drop the aging factor, the $1/t$ on the right-hand side of the rate equation for forgetting. If the aging factor is the only reason to regard the power law as being objectionable, then just getting rid of it might be a good strategy. Dropping the $1/t$, we get a new and improved rate equation:

$$\Delta A/\Delta t = -b^*A,$$

where the letter A has replaced M to make it clear we are considering decay of activation as opposed to forgetting of memory content. Although this equation superficially resembles the power law equation, it leads the investigation in an entirely new direction. In the first place it introduces something new into the discussion, a term that may be used to measure time passage – a timescale. To see how this comes about we will pay attention to units of measurement, not something that is generally important in psychology, but is important here.

A good entry point into a discussion of units is the power law rate equation. Starting with the units of the left-hand side we encounter a forgetting rate that has the dimensions of content/time. The units of content are not fixed and will vary according to whatever it is that people are forgetting. Content might be words, syllables, letters, numbers – whatever content is learned in memory tests. So, a forgetting rate might be words/second if participants are learning word lists. The right-hand side of the power law equation is b^*M/t, and it also must have the units of content/time. The M/t part takes care of the units, leaving no units for b. This means that b is dimensionless, a number on the real number line, which is why it can be the power in the power law.

Now consider the units for the new and improved activation law. The left-hand side has the dimension of activation/time which, admittedly, is a little confusing. Because activation is an abstract construct, it is not clear what units to assign to it. We can simply avoid this problem by creating a new left-hand side. Collecting all the activation (A) terms on the left-hand side gives:

$$\Delta A/A = -b^*\Delta t.$$

Now regardless of how A is measured, the left-hand side has those units dividing out. The left-hand side is now dimensionless, implying that the right-hand side must also be dimensionless. This means that b cannot be a power because it now has the dimensions of 1/time. It will make more sense if we use a symbol that does look like time, and the symbol usually reserved for this purpose is tau, τ. For the purposes of understanding how activation decay without aging works, it will help to write this expression in two ways:

$$\text{balanced form: } \Delta A/A = -\Delta t/\tau$$

or making the expression look like a rate equation:

$$\text{rate form: } \Delta A/\Delta t = -A/\tau$$

The solution to either of these equations is the negative exponential,

$$A(t) = A(0)\, e^{-t/\tau}.$$

The negative exponential also involves a power, but it works differently than a power law. Instead of time being the base for a power b, the power is time, t, divided by a timescale, τ. This division makes the power dimensionless, which it must be to make sense. The base is a number, the Euler number, which is universally written as e and which has a value approximated by 2.718. Entire books have been written about e, but here it suffices to just note that when nature describes decay, it uses e. The base could be any number, and we could rewrite the exponential with a base of 2 or 10. But this would lead to some extraneous terms, and it is not the way nature writes. Nature writes with e.

In the expression for the negative exponential, time, t, appears only in the context of its being divided by τ. This division has an important function; it makes τ the natural unit for measuring the flow of time. Natural, in this context, means that when the flow of time, clock time, is measured in units of τ, every exponential decay process follows the exact same curve. With the passage of a time interval of length τ, every exponentially decaying system drops to $1/e$, or roughly one third of its starting

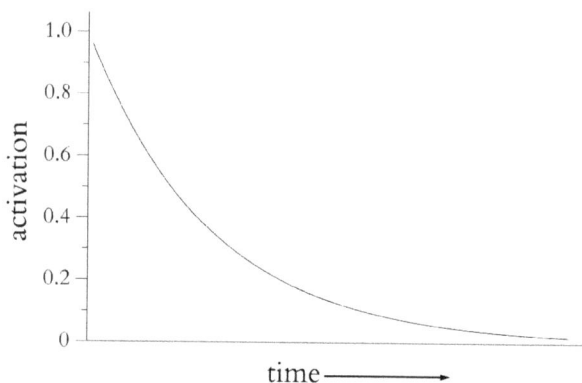

Figure 6.3

value. And with the passage of another time interval equal to τ, there is another reduction of $1/e$, and so on. Where exponentially decaying systems are individuated is in their values of τ. Every decay process has a specific value of τ and the larger it is, the more slowly the decay process runs. In nature, values of τ range from fractions of seconds to many times the age of the universe.

The shape of the negative exponential for $A(0) = 1$ is shown in Figure 6.3. From a purely visual perspective there is not much to distinguish the power law from the exponential. This is the root cause for why it is generally hopeless to develop a preference for any formulation of decay based just on shape and the fitting of data. A preference must be based on a principle, and here the principle is that although memories might have ages, activation does not.

The way that a negative exponential works is most clear in the balanced form of the rate equation. On the left-hand side is $\Delta A/A$, which is interpreted as the percent lost over some amount of time. That amount is given by the right-hand side, $-\Delta t/\tau$. The interesting thing about this is that there is nothing that stipulates when the time interval, Δt, occurred – right away after the system was activated or very late when the activation is mostly gone or whenever. Regardless of when that time interval happens it leads to the same percent loss. This is partly what it means to have decay without assigning ages.

A glance at a few physical systems that decay as negative exponentials will clarify how things operate when age is not a factor. Radioactive decay is one example. Radioactive substances have the property that the probability of decay over a given time interval is fixed and does not depend on how long the substance has been around. This may be unintuitive as people generally look at the historical record when thinking about what is probable in the future. But nuclei undergoing radioactive decay proceed as if they have no history. Nuclei that do not decay early on do not become resistant to decay in the sense that memories that survive become resistant to forgetting. There is no memory consolidation in radioactive decay. A decay event is never due, and it cannot be predicted. The time independence of decay probability entails that the number of nuclei that decay over a given interval of time depends only on the number of nuclei available for decay. This is why radioactive decay follows the negative exponential.

A second example is spatial, but the discounting of history takes the same form. Consider a lighthouse shining a beam into a layer of fog. A ship out at sea will see that light with considerably reduced intensity. This is because the individual photons in the beam scatter off water molecules in random directions. The lighthouse will be visible to the extent that some photons make it through the fog without being scattered. If the fog layer has uniform density, photons will have a fixed probability of scattering regardless of how far into the layer they have traveled. In other words, the travel history of a photon has no relevance to a scattering probability. Having traveled half-way into the fog, the photon is not *due* to be scattered. The travel history, how far a photon has penetrated the fog, is a way of dating the photon, giving it an age. Expressed in this way, a photon scattering in fog is another example of age not mattering. When age is not a factor, loss rates are proportional only to the amount that is present to lose, the prescription for the negative exponential.

The Meaning of Tau

The two rate equations being considered for activation decay have one thing in common: they both describe a drainage process that takes a full state into an increasingly empty one. They also look similar enough that in a fitting competition, it would be difficult to make a case for either being a clear winner. They have, however, divergent theoretical implications and offerings. The power law comes from a rate equation that introduces age as a factor in the decay rate, and this leads to an offering of a power, *b*. The exponential, in contrast, comes from a rate equation where age is not a factor, and this leads to an offering of τ, a timescale. If activation were the sort of thing that could be stored and consolidated, then we would be on a path to understand what sets values of the power law exponent, *b*. Here, however, we are considering activation as a form of transient energy, and this leads to consideration of what sets values of the activation lifetime, τ.

Physical scales always have two levels of interpretation, as numbers and as constructions. In everyday life the number aspect of τ is of paramount importance. People do not live in time measured in units of τ but rather live in time measured in seconds, days, and years, and it is often quite relevant how long something that is decaying can be expected to be around. But there is more to τ than being a number. At a level below the concept of number is architecture. τ, at the deepest level, is a construction that is formed from the interactions that describe the system where τ makes its appearance. There will, consequently, be many ways of constructing a τ. The τ in radioactive decay, for example, reflects the quantum physics that describes the transition probabilities of unstable nuclei. The τ in Newton's Law of Cooling reflects the factors that influence heat transfer such as heat capacity and surface area. The τ that appears in the attenuation of light traveling through fog reflects that physics of light scattering. Although these examples are physical, there are also many non-physical systems that are described by exponential decay, and these will also have τ constructions. Epidemiology, economics, and population biology are examples of just a few fields where negative exponentials might also be found. To this collection is now added activation decay in grouping and priming processes. As every τ has a construction, it is now appropriate and inevitable to inquire into the τ construction for activation decay.

The Phenomenology of Tau

A good place to begin to think about the construction of the τ that appears in activation decay is to step back and think about what it is that this τ does. This τ is an activation lifetime and so it locates the position of the phase transition in real time. From an engineering point of view, what τ achieves is the opening and closing of an information valve. On the short side of the phase transition the valve is open, and events come into relation with each other. It is through the experience of groups "being other than the sum of the parts" that useful information is picked up. On the long side of the phase transition the valve closes and the failure of relationships to

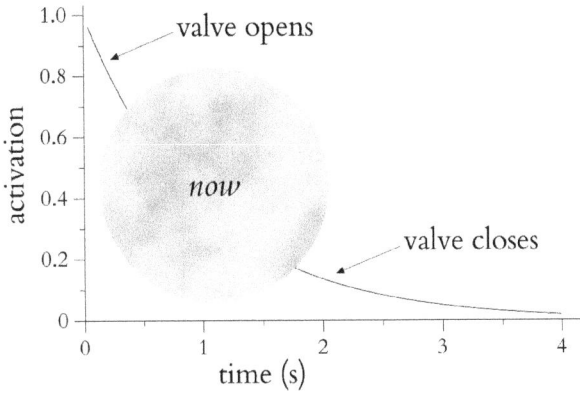

Figure 6.4

form leads to a succession of one thing happening after another. In this way τ acts as a throttle on emergence, setting the limits on our ability to extract information from the world. Before the valve closes there is emergence, and so it is before the valve closes where we live. Where we live is a pretty good way of thinking about the experience of the present moment, of *now*. This picture of temporality deserves a picture on the page, and it is given in Figure 6.4. We can now see exactly what τ does. It is the device that nature uses to temporally situate us.

This perspective is quite distant from the view of time as a kind of sensation in Stevens' law. We now have a way of thinking about time that is centered on how relationships are experienced and how the opportunity for experiencing relationships is set by this new thing, a timescale, τ. However, the understanding that activation-τ sets the horizons of our temporal experience does not give us its parameters, the physics and biology upon which it is composed. A yet larger principle is needed. There is a larger principle available, and it is arrived at by asking the question: Who benefits from the mind being temporally situated? Who benefits is always the most important question. Obviously, it is the body that benefits.

Note

1. A formal measure of the complexity of an object or pattern is the length of a message that completely describes it. So, although a mathematical expression may seem complex, there is nothing more complex than a data set. To see this

imagine telling somebody about the data, in such detail that they could reproduce them. Now imagine telling somebody about the mathematical expression, in such detail that they could write it down. The first communication will take a long, long time. It will be lengthy and wordy because every piece of data has a unique component, the noise in the signal. The second communication will take just seconds. It is inevitable that the message length of the data will vastly exceed the message length of the model.

CHAPTER 7

The Body in Time

An equation for human temporality does not exist, but if it did, it might look like

$$\tau = f(body),$$

a relation stating that out of the body comes the activation lifetime that sets the temporal horizon for the mind (or it might look like $t = f(self)$ a novel relation introduced by Wittman (2016)). At this point *f(body)* is a little impenetrable and more of a vision than a bit of mathematics. Clearly, if we are going to construct the activation lifetime from elements of the body, we will need some help in conceptualizing how this might be done. Arguably the path forward is not going to come from neuroscience. At issue here is not the neural response to a bar of light or a pure tone. We are not discussing something that might be sorted out in terms of receptors or receptive fields. Neither are we discussing something that might be sorted out by following blood flow in the brain. Rather, we are thinking about the totality of information that flows from time-based groups and the totality of ways in which the environment primes our thoughts and actions. The construction of τ is a bit mind-boggling simply because of the range of its influence.

The first issue to be confronted is finding a mode of discourse that is appropriate for describing the body's role in setting the activation lifetime. In seeking a point of entry into the unpacking of *f(body)*, we are drawn to the studies of animal scaling that were first initiated in the early twentieth century. In the physics and biology of animal scaling there seems to be an opportunity to match the level of abstraction of what τ is with a highly abstract description of the body. That allometric laws are highly simplified descriptions is not viewed here as a detraction. Rather it seems that scaling analyses benefit from taking a bird's-eye view of what it means to be an animal. But there is an aesthetic dimension to allometry as well. Scaling

103

relations are simple, and they hold the promise of pointing to a deep understanding of *f(body)* without requiring a deep understanding of how the brain works. There may be other ways to think about *f(body)*, but this path will turn out to be productive.

An Introduction to Allometry

Allometry is a branch of biology that relates body size to various aspects of animal life. (It also has wide application in botany, but here the focus will be exclusively on animals.) Allometric laws are generally power laws of animal mass, M:

$$\text{animal property} \sim M^b.$$

The animal properties that have demonstrated mass scaling fall into three broad classes: morphology – the shapes of animal bodies; physiology – the composition and structure of animal bodies; and behavior – the movement of animal bodies. Because animals develop, there are naturally two types of allometric laws. Ontogenetic allometries describe how animal properties change as the body increases in size and mass during development. Static allometries describe how animal properties are set by the mass of adults. The descriptive breadth of allometry is especially impressive in view of how reductive allometry is. That so much of animal life is governed simply by animal size is remarkable.[1]

Allometry is a thoroughly empirical branch of biology. Animal properties are observed, masses are observed, and then there is a regression analysis where the exponent b that gives the best fit of M^b to the animal property data is found. The interpretation of this relationship is that animals will have more or less of the property in question depending on their mass, with the steepness of the dependence being given by the power law exponent, b. Insofar as allometric laws are not really laws but fits to data, the regression techniques that are employed to find the exponent b are quite consequential. It is not uncommon to have competing exponents arising from competing statistical models. In this respect regression analysis in allometry is much like regression analysis in psychology. But unlike psychology, regardless of how the mass exponents are arrived at, they are taken quite seriously as organizing principles, as portals into ecology, animal form, and animal behavior. This is a bridging discipline that occupies the intersection of field observation, statistical analysis, and theoretical biology.

To get a sense of how scaling of animal bodies leads to power law relations, a good place to start is with the notion of a characteristic length, l. A characteristic length is not necessarily an animals' height or width. It is a little more abstract and better defined indirectly as:

$$l = V/S,$$

where V is the volume of the animal and S is its surface area. The characteristic length, l, is a kind of average length, and for a person would be smaller than their height, but larger than their torso diameter. Defining l in this way, the volume and surface area are given by the following power laws:

$$V \sim l^3 \text{ and } S = l^2.$$

These power laws are not discoveries, and they do not reflect anything deeper than the fact that animals are three-dimensional. In these whole-body analyses, the density of tissue is considered to be uniform, and consequently, there is no important distinction between scaling of mass, M, and scaling of volume, V.

$$M \sim V \sim l^3.$$

Putting all these relations together we arrive at a geometric relation between mass and surface area:

$$S \sim M^{2/3}.$$

This innocent relation contains little information beyond areas being two-dimensional and volumes being three-dimensional, but it will turn out to have an interpretation that makes it quite relevant to activation decay.

A scaling law that is derived from just geometry is not regarded as an allometry. The "allo" in allometry refers to the Greek word "allos" meaning *other*. In this context, *other* means other than just geometry. Staying with Greek roots, scaling laws that are not "other" must be "like" and so they are referred to as isometries. A foundational example of a scaling law that is *not* an allometry but an isometry is the 2/3 law for surface area. Implicit in the allo/iso distinction is that geometric scaling laws are treated as a kind of null hypothesis, what would be the case were there no additional considerations. In biology there may well be additional considerations, such as adaption, that will create scaling relations that deviate from isometry. In the practice of allometry, then, there is a regression analysis on mass that yields a mass exponent. If the exponent produces an isometry, then the investigation is complete, there is nothing further to be said. But if the exponent produces an allometry, the animal property is evidently organized by some aspect of physics or biology that may be worth investigating further. This decision, as it is based on regression, will depend on any number of statistical considerations, and most likely will be fraught and controversial.

Kleiber's Law

There may no better introduction to the spirit and intention of allometry than the controversy that has centered on the scaling of basal metabolism rate in mammals. As allometry will be playing a large role in understanding the activation lifetime, a little bit of time spent on the history of allometry will not be out of place and will provide important and necessary context for what follows.

In 1932 Max Kleiber published an article entitled "Body size and metabolism" (Kleiber, 1932) in which data were presented that supported a 3/4 law relating mass and basal metabolic rate (BMR):

$$BMR \sim M^{3/4}.$$

This relation, known as Kleiber's law, is famous first because it does seem to describe basal metabolism across an enormous range of animal sizes, but also because the 3/4 exponent is unexpected and unexplained. As surface area scales as $M^{2/3}$, if the exponent had been found to be 2/3 the interpretation would be that basal metabolism is a homeostatic mechanism that offsets radiative cooling at the animal's surface (an interpretation that would only hold for homeotherms – animals that maintain constant body temperature). Here is an example where an animal property might have been explained by simple geometry. The exponent, however, was measured to be 3/4 and this exponent does not have a simple geometric interpretation. There is still considerable uncertainty about how Kleiber's law should be regarded, and it is not clear whether any exponent is adequate to explain mass variation in metabolism. Allometry is an extremely reductive description and not every animal property can be reduced to a power law of size. However, when this discussion turns its focus toward human metabolism in a later chapter, we will find that there is an exponent and it is quite close to 2/3, the exponent for surface area.

Bringing Allometry into the Animal Mind

The animal properties that have been the focus of allometry – morphology, physiology, and locomotion – do not exhaust the scope of animal life. These properties reflect the materialistic aspects of being an animal and animals are not just material. All animals have sensory systems, and all animals perceive a world. Perceptual systems serve one principal function, the transformation of energy on sensory surfaces into information that has utility within an ecological niche. At root, all animals pick up information, and it is within the act of information pick-up that psychological allometries might be contemplated.

Different animals pick up information in different ways with wildly different sensitivities, but there is a class of animals, mammals, that live in a world composed of spatial objects and temporal events. This is an important observation, as it is the case that the pathway to objects and events is perceptual organization, Gestalt. Without organization the perceived world must consist only of simple structures in light. A good introduction to a world that is not populated by objects and events is provided by considering the frog visual system (Lettvin, Maturana, McCollough, & Pitts, 1959). Frogs see exactly four structures in light: moving edges, stationary edges, overall dimming, and the movement of small fly-shaped things. The key point here is that frogs do not see those fly-shaped things if they are not moving because frogs do not perceive a world of objects, and a fly is an object. In contrast, consider a dog that picks up a thrown ball, returns with it, drops it, backs away, and then looks up to await another throw. This animal is demonstrating an appreciation of both object and event structure. To be sure, viewing the world in terms of objects and events does not entail any of the accouterments of higher-order cognition. A mind that creates groups in time and space does not necessarily engage in tool use, gaze following, the recognition of other minds, the recognition of object permanence, or appreciating the concept of irony. But it does entail a mind with a particular form of temporality.

All animals that do Gestalt must contend with a world that is continuously in flux, and that it is in flux mandates that locality or proximity must be a strong determinant of what is grouped together. If an animal can perceive event structure, then it must have a temporality that admits two phases: grouping and succession. The proposition is that animals that perceive events also have a capacity of apprehension, a phase transition between group formation and succession. Whether or not this is a comfortable idea, the theory where phase transitions are created by activation lifetimes need not be limited to humans. Humans, in fact, have no priority in the universe of object-perceiving animals. Consequently, it makes sense to ask whether τ may be generalized to object and scene perceivers generally, or mammals at least. Once it is allowed that mammals have an activation lifetime setting their phase transitions, there is motivation to examine whether τ = *f(body)* has body-size scaling. In this picture mammals would share a common temporal psychology in much the same way they share common homeostatic mechanisms.

The general situation regarding animal timescales is that the larger the animal, the larger the value of the timescale, regardless of what that timescale measures. In mammalian physiology, for example, heart period, respiration period, and blood circulation time all increase with increasing mass. Similarly, timescales generated by locomotion also increase with animal size. A walking mouse will traverse a distance equal to its body size much more quickly than will an elephant. If τ respects at least the proportionality of bigger body–bigger time, then larger animals would have longer decay lifetimes and smaller animals shorter decay lifetimes. As the decay lifetime sets where the phase transition in grouping occurs, the implication is that body size would impact where succession occurs. Time separations between neighboring events that would lead to succession in smaller animals might still permit grouping in larger animals. Another way of saying this is that larger animals would have a bigger present moment, a bigger *now*. This idea is speculatively depicted in Figure 7.1 as a linear relation where larger animals have larger spans of apprehension and so longer durations that mark the point of phase transition. If any of this is true, then perhaps very large animals might have a phase transition at 3±1 s and very small animals at 1±0.5 s.

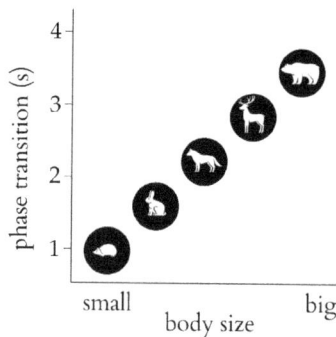

Figure 7.1

The picture presented in Figure 7.1 is, in fact, nothing more than a picture. We are not aware of any measurements of phase transitions in animal cognition. And consequently, there are no published allometries of phase transitions. Indeed, there are very few published measurements of phase transitions in people. An oblique approach is required here, one where phase transitions are inferred rather than directly measured (as when Fraisse estimates when *bit* followed by the sound of *ter* stops sounding like *bitter*). Here the work of Margaret Schleidt and her colleagues becomes relevant. Schleidt was able to effectively estimate the phase transition in time-based grouping, not by studying grouping, but by analyzing the choreography of human movement patterns. Her studies of human ethology find their relevance to animal cognition through the influence they had in motivating a subsequent investigation of mammalian movement patterns. Although a single study, an opportunity to infer activation lifetimes for animals in the zoo is so improbable that it is irresistible.

Ethology of Human Movement

The idea that within the record of the motion of the body is written the temporality of the mind is profound. But it is also problematic. This is a record that is not simple or straightforward to read. It requires technique. There are techniques for reading human behavior and there are different techniques for reading mammalian behavior. The humans are simpler to understand, and it is best to start with them.

Schleidt's technique begins with the recognition that human behavior is hierarchically organized, and that each level of the hierarchy is defined by its intentions. For example, at the top level of the hierarchy, the intention may be quite broad, such as *make dinner*. This intention will lead to subsidiary goals and subsidiary intentions such as *prepare vegetables* or *set table*. At the bottom of the hierarchy is kinematics, the physical trajectories of body parts over time. The hierarchy level where Schleidt measures durations is, in a very specific and logical sense, one rung up from the raw kinematics. One rung up from the kinematics is where sets or groups of trajectories are formed. In this Gestalt way of thinking about body motion, the trajectories are the parts of groups, and the groups are where behavior becomes intentional. These low-level groupings are referred to by Schleidt as *action units*.

Some examples may clarify what an action unit act is. If, for example, a person wishes to indicate assent, they nod their head up and down. If they wish to indicate dissent, they shake their head side to side. These gestures involve maybe three or four motions of the head. The individual motions are at the bottom of the hierarchy and are not of interest because they are just the parts of a group which is forming. The gesture, where head motions are grouped into a meaningful event, defines the level at which action units emerge. This definition places the action unit at the lowest level of the action hierarchy where an intention is expressed. In this way of thinking, the kinematics of head movement are not intended but are merely the vehicle for the expression of an intention. Similarly, consider a person who intends to scratch an itch. They do this with a few back-and-forth hand motions. The individual hand motions are *not* action units. The action unit is created when the motions are collected into a group that constitutes a scratching event. Again, a person does not intend hand motions, they intend the group because the group is where the meaning of the activity resides.

Schleidt and her colleagues view human behavior as a stream of action units. This stream is defined by two independent characteristics: what the acts consist of, their purposes and goals, and how long they last – their durations. Schleidt's methodology involved categorizing the stream based on what people are doing (grooming, working, and playing, for example), and then forming duration histograms within each category. The principal finding was that durations within and across categories (and cultures) are typically 2±1 s (Feldhütter, Schleidt, & Eibl-Eibesfeldt, 1990). This is by now a familiar quantity. Apparently, the durations of action units are also the durations that set the phase transition in time-based groups. It is surely not a coincidence that *bit* and *ter* go into succession when separated across the time span that it takes to nod assent. This convergence has not gone unnoticed, but it has also not received a satisfactory explanation.

In thinking about the convergence of nodding and *bitter*, it seems that there is an important distinction between what is volitional and what is mandated. Behavioral acts and their durations are volitional. If a person wants to, say, nod their head for 10 seconds, there is nothing preventing that. In contrast, the temporal constraints on time-based grouping are not transactional and are not freely entered. If people want to play music, for example, they must choose a tempo faster than 30 bpm because it is not possible to generate rhythmic pulse when 2 s separates successive beats. The suggestion here is that the phase transition at 2±1 s that disrupts groups is a property of mind whereas the 2±1 s that measures the duration of action units reflects the shaping of free behavior. The implication is that behavioral acts, because they are free and therefore tunable, are entrained to the capacity of apprehension, which is not free and not tunable. In this way behavioral acts are shaped so that they have beginnings and endings that fit within Fraisse's capacity of apprehension.

This raises the question of why we accept this tuning, why we behave so that beginnings and endings fit into 2±1 s. Presumably it is so that beginnings and endings are perceived in relation to each other. The quality of being perceived in relation is necessary if we are to have the sense that endings end what beginnings begin. So, although it is always possible to string together an arbitrarily large behavioral act, if the two moments of time, *beginning* and *ending*, are to feel connected and be perceived in relation to one another, the *ending* moment must occur within the activation decay period that was initiated by the *beginning* moment. When beginnings and endings are in a common group, the behavioral act will fit into 2±1 s.

The upshot of behavioral entrainment is that the location of the phase transition in grouping, and correspondingly an estimate of the activation lifetime, τ, may be inferred from the durations of action units. This is a methodological serendipity. It means that an observer can penetrate the subjective experience of another by breaking down their overt behavior. When we think about nonhuman mammals this methodological finesse may be the only opportunity for estimating whatever passes for their 2±1 s phase transitions. It would be wonderful if a rat or a dog could tell us at what separations two ticks sound like tick-tock, but until that happens, we will be relying on indirect estimates of their phase transitions.

Ethology of Mammalian Movement Patterns

The extension of Schleidt's techniques to mammals generally is problematic but was attempted by Gerstner and Goldberg (1994). The principal difficulty is that people lack common ground with other animals. The foremost methodological problem in Schleidt's ethology was developing a meaningful and reliable way to segment a continuous record of human behavior into action units. The segmentation problem was largely solved by the circumstance that people observing other people recognize any number of subtle movements that signify beginnings and endings. This knowledge is tacit. Rather than being rule-based, it is of the form of *you-know-it-when-you-see-it*. It is the existence of this tacit knowledge that obviates the need for step-by-step instructions from Schleidt and colleagues on how to parse extended activity into an action unit stream. As nonhuman animals do not have human minds, partitioning their kinematic motions into meaningful groups is inevitably approached without the familiarity afforded by watching videotape of members of our species.

The action segmentation problem in mammalian ethology is addressed by first assembling an exhaustive set of movement patterns into a library known as an ethogram. Extended videotaped recordings of animals doing what they do are organized by placing every moment of animal behavior into at least one movement pattern contained in the library (animals may be doing many things at once with their various body parts). The intention of constructing an ethogram is to create an objective record of animal behavior – a record that does not rely upon tacit knowledge. *Objective* is always a worrisome term and here it simply means that both the construction of the library and the assignment of patterns within the library are subject to assessments of inter-rater reliability. That is, different people must at least agree on the descriptions of animal movement patterns. Where objectivity falls short is in using the ethogram to measure pattern duration. The problem of how to partition repetitive motion is not solved by the ethogram. Sequences such as chewing and tail wagging will inevitably contain pauses, and the decision about whether a given pause marks an ending is intrinsically difficult to make. Schleidt was able to use her tacit understandings of human behavior to recognize whether she was observing, say, three short and separate scratching events or one very long scratching event. Animal minds and intentions cannot be penetrated through the shared common ground that common species membership confers. When it comes to other animals we are, in a very real sense, outside observers.

Gerstner and Goldberg (1994) compiled a massive data set of movement pattern durations from an ethogram formed on six mammalian species: giraffe, okapi, red panda, raccoon, roe deer, and kangaroo. The ethogram in this study included 113 movement patterns and was quite detailed. A sense of the level of detail is given by the entries having to do with lying still: "lying still, eyes open," "lying still, eyes closed," and "lying still, eye status unknown." Unlike Schleidt's studies, which focused on specific categories of behavior such as working and grooming, this study included whatever behavior was evident during the recording window. Because the behaviors in the library were not culled in any way, durations span an enormous range, from the briefest subsecond hiccup to extended episodes of ruminating and standing with eyes open. Consequently, the resultant duration distributions of movement patterns are not as strongly peaked as those created by Schleidt for action units. Although it may not be entirely appropriate to draw inferences about characteristic values from broad and diffuse distributions, these data are unique and so an analysis of central tendency is justifiable – at least to see where it leads.

The movement pattern duration means and standard errors as reported by Gerstner and Goldberg were (all values in seconds):

Giraffes:	4.82 ± 4.99
Okapis:	3.42 ± 4.06
Roe deer:	2.70 ± 3.00
Kangaroos:	4.04 ± 6.04
Red pandas:	3.76 ± 5.13
Raccoons:	1.62 ± 1.65

As the standard error is used in statistics to indicate the precision with which the mean has been measured, it is unfortunate that the standard errors are larger than the mean values. The large errors are testimony to just how diffuse the duration distributions are. Also of concern is that six mean values for six mammalian species do not comprise a rich data set for a regression analysis. Nevertheless, the circumstance that these data are unique in the literature requires them to be taken seriously. What follows is a first attempt at developing an allometric law for the durations of mammalian movement patterns.

Figure 7.2 plots mean movement pattern durations and animal masses[2] for the six species analyzed by Gerstner and Goldberg. Because allometries are generally framed as power laws, it is traditional to plot them in the log–log plane where they are transformed into straight lines. Although the transformation may be potentially confusing, there is no simpler data structure than a straight line and straight lines lend themselves to visual understanding.

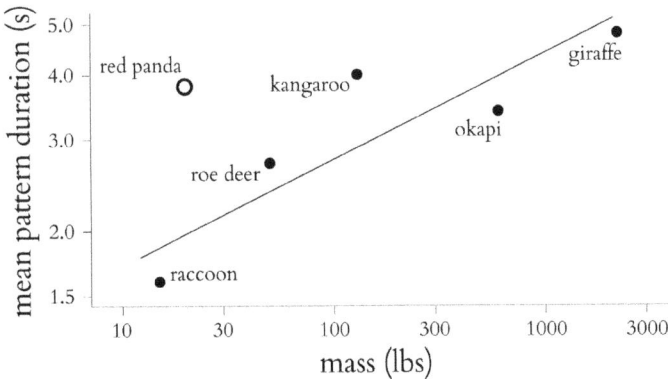

Figure 7.2

There are two ways of visually organizing this plot. One way is to regard the raccoons as engaging in unusually short movement patterns. If the raccoons are removed from this data set, animal mass and movement pattern duration would appear to be mostly unrelated. The lack of relation would suggest that there is a characteristic duration for mammalian movement patterns of about 3–4 s. This conclusion would be quite consistent with what Schleidt found for human behavior and is also the conclusion that Gerstner and Goldberg drew. A different conclusion is reached if instead the red pandas[3] are viewed as being unusually slow-moving given their relatively small mass. In this regard it may be noteworthy that the red pandas were the only nocturnal mammal included in the study and videotaping sessions generally ended in the early evening. If we exclude the red pandas there is now a clear linear trend where increasing mass is associated with longer movement pattern durations. Such are the perils of interpreting six data points. Nevertheless, this trend invites the fitting of a line that will lead to an allometric law. An allometry of the duration of movement patterns is formally expressed as:

$$\text{pattern duration} = a\,M^{b},$$

where M is animal mass and b is an unknown power that becomes known through a regression analysis. The regression analysis begins with taking the logarithm of both sides. The logarithm turns the power law into a linear relation:

$$\log(\text{pattern duration}) = \log(a) + b\log(M).$$

Cast in this form, the intercept, $\log(a)$, and power, b, are easily found using standard procedures for finding a line that best cuts through the data.

The regression line shown in Figure 7.2 does not include the red panda (the open circle). The remaining five species have log masses and log durations that are highly correlated, $r = 0.88$. This is a large correlation, although surely a poor estimate of the true correlation given that it is based on highly diffuse duration distributions from just five species. Still, the probability that the true correlation is zero and the data simply reflect the randomness of chance is 1 in 40; $p = .025$.

One issue that must arise in interpreting mammalian movement pattern durations is that some events in the ethogram are constrained by physics whereas others are not. For example, some ethogram entries refer to motions of the whole body such as postural adjustments whereas others refer just to the motions of small body parts, such as the tongue used in grooming and chewing. Where the whole body is involved, it might be expected, for purely physical reasons, that the time to move would be highly correlated with mass. Bigger animals have bigger limbs and more mass to move than smaller animals. Where physical constraints are active, there will always be body-size scaling of the isometric–geometric variety. Most events in the ethogram, however, did not involve gross postural adjustments or other whole-body motions. If the correlation is real and if it is not trivially explained by body inertia, then it becomes interesting because then the animal mind becomes relevant.

The regression model shown in Figure 7.2 produced $b = 0.19 \pm 0.06$. If this model was reporting on some property within the scope of biology, perhaps dealing with some aspect of morphology or physiology, b would be interpreted in terms of physical quantities, material aspects of the body and the environment in which the body resides. A mean movement pattern duration, however, is of a different order entirely. Although pattern durations are inferred from the movement of body parts, the duration itself is considered to signify a property of mind, the span of an animal's sense of the present moment. At least this is the interpretation given to pattern durations by Gerstner and Goldberg – the enormous effort required to create the movement pattern library was justified by the prospect of studying animal minds. An allometry of a property of mind is not something previously contemplated in the published literature, and so caution is invited. Recognizing that an allometry derived from just five data points is not reliable, before committing to an interpretation of $b = 0.19 \pm 0.06$, it might be prudent to find a paradigm that leads to a richer data set. The ethology of mammalian behavior is brutally difficult, and it is so much easier to do to human psychophysics. The richer data set will be found in human rhythmic behavior.

Perspective on Bringing Allometry into the Animal Mind

Ethology is difficult to do, but it offers the promise of a view of animal temporality that is free from experimental designs and other forms of human interference. There is something enchanting about observing the entire repertoire of behavior and not focusing on just responses to stimuli. And the idea of using the durations of movement patterns to inform on the placement of phase transitions seems quite elegant. There are, however, some conceptual issues that need to be faced. Besides the technical problems concerning the parsing of extended behavioral streams, there is the more fundamental issue of what exactly mammalian movement patterns represent. In Schleidt's ethology, the action unit was identified as the first level in which the kinematics of body parts cohered into groups that had identifiable intentions. There is no question that other mammals do have intentions, but the practical identification of movement patterns did not involve recognition of groups or intentions. In place of the kind of understanding that comes from observing members of your own species, there was the rote application of rules for defining beginnings and endings of motion sequences. In short, action units in Scheldt's catalogs reflect meanings and context, whereas the movement patterns of mammals reflect the application of an explicit rule structure that refers just to kinematics. This is not a small distinction, and there is simply a lack of clarity about what the duration of a mammalian movement pattern signifies.

Nevertheless, it is arguable that something positive has been achieved in the analysis of mammalian behavior. Two options have been presented for the interpretation of mean pattern durations. The first option is that there is a characteristic value of about 3 s that governs the sequencing of movement patterns. If this is true, then there is an interesting convergence between humans and other mammals in how body motion is choreographed. The second option is that there is not a characteristic value of movement pattern duration, but rather an allometry that follows the usual trend that larger animals are associated with longer time intervals. If the allometry is real and if the durations of movement patterns reflect proximity constraints in scene analysis, then this allometry may inform on how the mammalian version of an activation lifetime, τ, scales. Regardless, this is as far as an analysis of *f(body)* can go without better data.

Notes

1. The reappearance of power laws following a discussion of the power law of forgetting may seem meaningful, but it is not. It is simply an illustration of two ways in which power laws may arise. Power laws in the description of forgetting curves derive from the resilience of old memories. Power laws in allometry, in contrast, are ultimately rooted in geometry.
2. Mass data were not given in Gerstner and Goldberg (1994). Typical masses for the various species were estimated using the provided data on gender, age, and subspecies.
3. Red pandas are not bears and so are not pandas in the familiar sense. Rather, they are genetically linked to raccoons even if they are apparently more dilatory than raccoons.

CHAPTER 8

An Allometry for Rhythmic Pulse

Gestalt psychology has historically relied upon the shared experience that being a person among other people affords. In the absence of theory there is just phenomena, and there may be nothing less theoretically understood than the being in relation that underlies perceived groups. Consequently, Gestalt psychology is, for the most part, an informal sharing of phenomena – an invitation to look or listen to something that emerges from being in relation. It is the field of cognitive psychology's acceptance of informality that gives Fraisse the license to simply list without citation a few examples that illustrate his concept of the capacity of apprehension. His readers, being people, can join along in saying *bit* and *ter* and experience that there are in fact two perceived states, grouping into *bitter* and succession, depending on the syllable separation. What is not clear is whether the informality of shared experience can lay the foundation for the rigorous system of measurement that is required for phase transition allometry. The 2±1 s language that has served the discussion up to this point will not work in a statistical analysis that leads to a power law of mass. Precisely locating phase transitions requires that there be clarity in identifying when a person is experiencing a time-based group and when they are not. Achieving that clarity is not trivial and certainly not something that lends itself to introspection. In place of introspection the enterprise of allometry must be based on finding behaviors that literally announce, "I am experiencing grouping" or "I am experiencing succession." The question is where these clearly indicating behaviors might be found.

Pulse: A Paradigm for the Measurement of a Phase Transition Allometry

An optimal environment for studying the phase transition between grouping and succession is the feeling of rhythmic pulse that occurs when successive beats are fused into a tactus – a beat train. What makes this

117

environment optimal is that the regimes of grouping and succession produce very different behaviors and, most importantly, the behaviors are overt and easily observed. When a person is drumming at a tempo where they can feel rhythmic pulse, their performance will be accurate in the sense that it is steady and on the beat. But when a person is attempting to drum at a tempo that is so slow that they lose pulse, their performance will be erratic and inaccurate in the sense that drum strikes will not follow the divisions in time that are prescribed by the tempo. The massive experiential difference between feeling pulse and feeling lost is a potential basis for developing an allometry for the phase transition between beat grouping and beat succession.

The experiment described here does not involve complicated techniques, but it will be easy to lose track of the thread amid the myriad issues that arise in the practical identification of the pulse/succession experiential boundary. At the outset of what will be a long discussion of drumming contour, it will help to have two images in mind. The first is that of activation overlap. This idea was introduced earlier, in Chapter 5, as a way of understanding how *bit* could group with *ter* to make *bitter*. Here we are interested in rhythmic pulse, but the template is the same. So, in Figure 8.1 the template is illustrated with two moments of drum strike instead of two moments of *bit* and *ter*. Each drum strike produces a history of activation decay and if there is history overlap then the conditions for the experience of rhythmic pulse are met.

The second, but no less important image to keep in mind, is how allometry in the decay lifetime, τ, leads to allometry in the experience of rhythmic pulse. This is taken up in Figure 8.2 where two possible decay

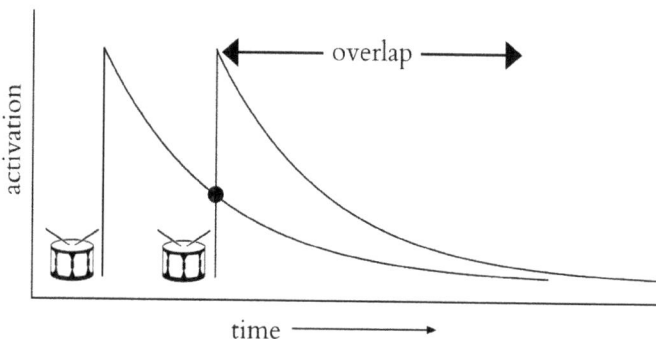

Figure 8.1

histories of a single drum strike are contemplated. One history, labeled *slow*, is produced by a person with a long decay lifetime. The other history, labeled *fast*, is produced by a person with a short decay lifetime. As timescales in animal behavior and physiology tend to scale positively with size, it would not be errant to associate the slow decay with a tall person and the fast decay with a short person. For the purposes of illustrating the argument, it will be assumed that both people have the same sense of rhythm and are identical in every way except for their activation lifetimes.

Implicit in the construction of the activation template is the idea that there is some minimum activation where the memory of a drum strike has decayed to the point where it is no longer available for overlap or grouping. So, for the purpose of illustration it will be supposed that in each person this level has a numerical value of 0.2. In Figure 8.2 this minimum level is depicted by a horizontal line intersecting the y-axis at an activation of 0.2. The key observation to be made is that the line of minimum activation cuts the two decay histories at two different points in real time. The short person reaches an activation level of 0.2 at about 0.8 s whereas the tall person reaches the 0.2 level at a time about 1.6 s. As we are thinking about these people as being drummers, 0.8 s is the maximum time between drum strikes that will produce a rhythmic response in the short person, and 1.6 s is the corresponding limiting duration in the tall person. Reframing these times in terms of tempo, the short person loses pulse and experiences

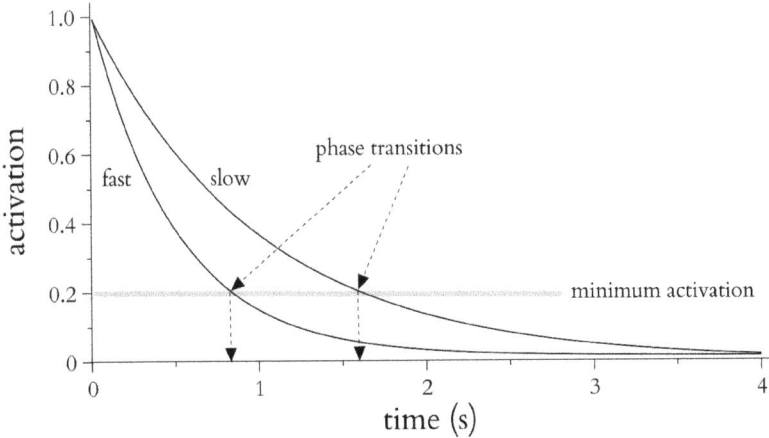

Figure 8.2

a phase transition at a tempo of about 75 bpm whereas the tall person has that experience at 38 bpm. As will be shown in the actual experiment, these numbers are not that far from reality. The image to keep in mind then is that variation in decay lifetime induced by body size will be reflected in the tempo where drummers lose their sense of rhythm.

Pictures of Rhythmic Pulse

There are few people, if any, that can steadily beat a drum at 20 bpm where 3 s intervenes between each beat.[1] Conversely most people can provide a steady drumming performance at 80 bpm where 3/4 s separates each beat. Somewhere between 20 bpm and 80 bpm there is some tempo where most people will find their limit – that tempo where their sense of rhythmic pulse begins to fail. That watershed tempo, and the temporal gap associated with it, serves as a measurement of that person's rhythmic phase transition. Conceptually, then, the way forward is clear – find watershed tempi. Practically, however, finding a watershed tempo is not trivial and involves understanding some sensory psychophysics as well as some of the theory of random processes. The following discussion will provide a practical introduction to the art of finding the watershed tempo in real drumming data. That there is art involved will become clear.

Drumming performances are for the ears, they are not for the eyes and are not intended to be visualized. As it is generally obvious if a musician is not playing in time, it might seem unnecessary to introduce spatial representations of what is fundamentally an acoustic signal. There are, however, ways of spatially representing drumming performances that take advantage of the visual system's exquisite sensitivity to landscape contour. There may be nothing that people do better than extract statistical information from landscapes. After all, it this capacity that allows us to evaluate whether the place where we intend to walk is, in fact, walkable. This section, then, consists of two tutorials. The first is on how drumming performance may be turned into spatial contour. The second is on how spatial drumming contours may be analyzed to infer when a person is feeling rhythmic pulse and when they are not. This second tutorial involves quite a bit of basic sensory psychophysics as it is critical to distinguish the state of being lost from the errors that will inevitably accompany slow drumming. Having laid this groundwork, an allometry in the pulse phase transition will not be difficult to produce.

A natural place to start thinking about how drumming might be visualized is the waveform of a drumming performance. Figure 8.3 displays an

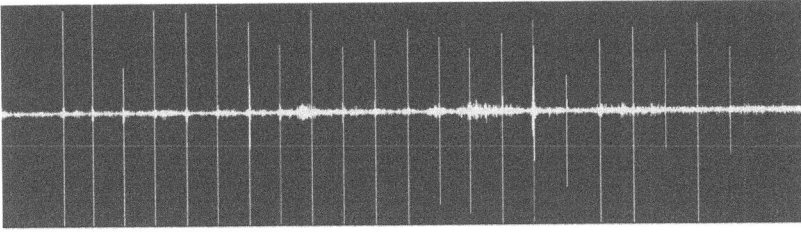

Figure 8.3

actual drumming waveform as recorded as an audio track by a digital-audio workstation (DAW). In this representation, time runs along the *x*-axis and the *y*-axis is the signal amplitude. The regularly spaced vertical lines depict the percussive sounds made by drum strikes. These lines are so thin that it might seem that what is being represented is just the moment of drum strike. In fact, the vertical line shows the entire acoustic signal of each strike. It is just that percussive drum strikes have extremely brief decay and rise times and so fail to be resolved at the magnification of Figure 8.3. Also visible is a horizontal line that is not perfectly uniform – the amplitude waxes and wanes. This component depicts fluctuations in the ambient room noise that was present during the recording.

A casual inspection of the waveform makes clear what the visual system is good at and what it struggles with. There are two salient aspects to the visual appearance of the series of drum strikes. One is that the drum strikes are grouped into a pattern. The waveform looks like something – it looks like a picket fence. This is Gestalt; the individual vertical lines are seen in relation to other lines, and the grouping percept is of a pattern that has the emergent property of looking periodic and regular. A second thing that might be noticed is that the vertical lines are not all the same size. The line tops and bottoms group into a contour that undulates up and down. This Gestalt is different from the fence Gestalt. In the contour Gestalt the tops are connected separately from the bottoms making two separate contours. So far so good, but there is additional information in the waveform that is quite difficult to see. This information is contained in the sequence of widths that separate pairs of neighboring vertical lines. These widths have a name, the inter-beat interval (IBI), and their constancy gives the picket fence the appearance of regularity. However, this performance was made by a human, and it is not perfectly regular. This is a key point; it is the deviations from regularity that stipulate the degree to which a drumming performance expresses pulse or expresses being lost. The problem with the audio track way of looking at drumming is essentially that the Gestalts that are vivid are not relevant to the detection of rhythmic pulse, whereas the inter-beat intervals, which are relevant, do not form an identifiable Gestalt – a contour. What is required here is a contour composed of inter-beat intervals. Such a contour requires a mathematical construction where the sequence of widths is transformed into a time series.

Time Series Representation of Rhythm

Time series are quite common in virtually all sciences with the notable exception of psychology. Any discipline that is interested in magnitude variation over time will employ time series. In economics, for example, there are time series of stock market closings, percentage rates, home prices, and so on. In meteorology there are time series of rainfall, high and low temperatures, snowpack, etc. The utility of plotting time series is that these graphs take advantage of something people are extremely good at – looking at and assessing contours. To be clear, the information that is present in a time series could be presented in a table, but the visual impact of a contour simply cannot be replicated in a table.

The first step in visualizing drumming accuracy and stability is to compute the time series of inter-beat intervals from the moments of drum strike. As an example, consider a brief drumming performance that consists of just three strikes that occurred at moments M_1, M_2, and M_3. From this brief performance there are 2 IBIs:

$$IBI_1 = M_2 - M_1, \text{ and } IBI_2 = M_3 - M_2.$$

The second step is to plot the time series in a format that takes advantage of the visual system's contour sensitivity. This was done in the preparation of Figure 8.4. The x-axis consists of the sequence, 1, 2, 3, . . ., which indexes the inter-beat intervals: IBI_1 is plotted first, IBI_2 is plotted second, and so on. The y-axis plots the magnitude of each IBI by assigning a vertical height that is proportional to the magnitude. In this way a drumming performance is turned into a landscape contour that takes full advantage of the attunements of the human visual system. The contours depicted in Figure 8.4 are time series formed from single performances at the target tempi of 15, 30, 40, 60, and 120 bpm. They all resemble rough contour, but with a little background in the psychophysics of error it will be possible to distinguish the error produced by rhythmic pulse from the error produced by being lost.

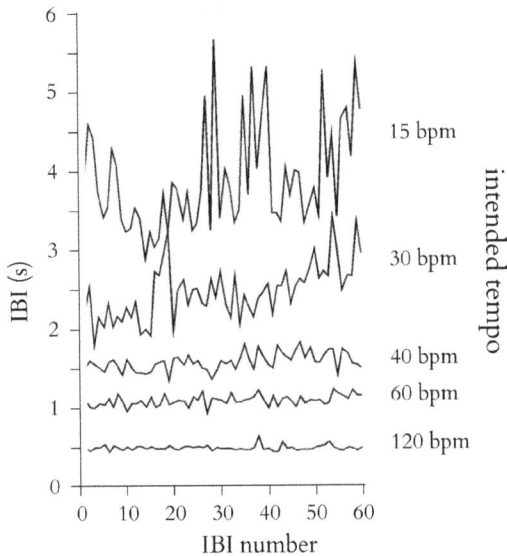

Figure 8.4 Copyright © 2022, American Psychological Association

The graph of an IBI time series when a person is experiencing a strong sense of rhythmic pulse will resemble a relatively flat landscape that is adorned with little hills and valleys. Flatness in the IBI time series arises when the intervals are the same size and so have the same heights. IBIs will be constant to the extent that the drummer feels rhythmic pulse and uses this form of tacit knowledge to strike the drum on the beat. While a drum machine would produce a time series that is completely flat, people are not machines, and their drum strikes will always be a little early or late – even when they have a strong sense of rhythm. In an IBI time-series contour, early strikes make dips and late strikes make bumps. When the early and late strikes are randomly intermixed, as they are in actual performances, the IBI time series contour will ripple. What can be expected then from a human drummer that is experiencing rhythmic pulse is a rippling contour that stays at a steady average height. In contrast, the graph of an IBI time series when a person is not feeling rhythm is not particularly well-specified. Experienced musicians will stop playing if they have lost the beat so, in a sense, the absence of a time series might be the truest expression of not feeling rhythmic pulse. Nevertheless, in an experiment that locates the pulse/succession phase transition, drummers will be required to make some sort of performance even when they would rather not. Such lost performances will not express a constant tempo, and the inter-beat intervals will not make a rippling landscape.

The full spectrum of rhythmic experience is captured by the collection of contours illustrated in Figure 8.4.[2] The contours span the gamut from flat with ripples to flat-ish with big ripples to wandering to erratic. This much is obvious and that is the point; spatial contour has impact for human vision. Being able to look at a performance and ponder its structure is vastly superior to and easier than attempting a structural analysis in real time while listening to a performance. That said, there is some subtlety to the visual analysis of drumming performances. To understand in a more precise way how to compare performances at different tempi will require some basic knowledge about how sensory systems (audition, vision, touch) work. We will be paying particular attention to the progression in error growth from the small ripples at 120 bpm to the big ripples at 40 bpm. This progression involves Weber's law, a topic first introduced in the first chapter in a discussion of how rats make errors in the judgment of time intervals.

Just-Noticeable Differences

A core property of perceptual systems is that they are sensitive to change and consequently a core determinant of a perceptual system is how much change must occur before a change is noticed. To introduce the constellation of ideas that arise in noticing change, let us start with the change that is experienced when prices rise by $1. If a cup of coffee that once cost $3 now costs $4, that will be noticed. However, if the price of a car that once cost $30,000 now costs $30,001, that would not be noticed – that difference is just too small to be a concern. A dollar is a fixed, raw, quantity but it is never perceived in the flesh, naked, as $1. A dollar can be both a lot of money and no money at all; it depends on the context in which it appears. The same thing is true of a normal speaking voice (it may be loud in a library but inaudible at a rock concert), the light of a star (bright in the night sky but invisible in daytime), and so on with virtually every dimension of human experience. Context in psychophysics is given formal definition as contrast, and contrast is given mathematical definition as a ratio. In the coffee context, the $1 is experienced as a 1/3 or 33 percent increase, whereas in the car context the $1 is experienced as a 1/30,000 or 0.0033 percent increase. The former fractional change is highly salient whereas the latter is not, it is in the noise. There is a lesson here.

Apparently, the way people notice change and experience sensations generally is negotiated through fractional differences – percentages. Large-enough fractional differences will be noticed. How large is large enough? That is the issue addressed by the concept of the just-noticeable difference, commonly known by the acronym JND.

The JND is how much must be added (or subtracted) from a given background so that a large-enough fractional change is produced that can be just barely noticed. Implicit in this definition is that JNDs are never constant and will be large or small depending on the background magnitude; a whisper might be a JND in a quiet library whereas shouting might be a JND at a concert. The following equation gives the formal definition of the JND:

$$\text{JND} = K^* \text{ background magnitude,}$$

where K is the fractional amount that must be added or subtracted from the background to be just barely noticed. Although this equation does define what a JND is, because human sensitivities operate in terms of percentages, its real importance is in introducing K.

This equation, taken literally, is simply a recipe for finding just-noticeable differences. The equation itself does not tell us anything specifically about K. It may well be the case that for a particular form of experience a 5 percent change is barely noticeable at low magnitudes, but that a 10 percent change is required to be barely noticeable at high magnitudes. There are, however, circumstances when K is truly constant and does not change when the background magnitude changes. In this case K is not just a practical measurement of what might be noticed, it is a quantity that characterizes the inner workings of the system that is capable of noticing. When K is a constant, the sensory system is referred to as being Weberian and K is referred to as a Weber fraction, honoring Ernst Weber, one of first people to quantitatively study perceived intensity.

It happens that most sensory experiences are Weberian across a range of background magnitudes. In consequence, it is possible to think about sensory experiences holistically as having a particular K – a particular level of sensitivity. Those things that people are quite sensitive to may have K values of just a few percent; shock is about 1 percent and brightness is about 2 percent. Slightly less sensitive is perceived weight, which requires a 5 percent addition to be noticed. There is also a K for the experience of rhythm and these ideas have immediate applicability to the problem of reading from a drumming time series whether a person is feeling rhythmic pulse or not.

Natural Error in Drumming Performance

The errors that people make are closely allied with what people can notice and experience as a deviation. Practically, to monitor and control the production of error, it is necessary that the magnitude of the error be noticeable, greater than a JND. It is not possible to monitor what is not noticed. This applies as well to placing beats at a steady tempo as it does to sensory experience and processes of judgment generally. Deviations from steady tempo would be expected to be at least as large as the JND for noticing that the tempo has strayed and become too fast or too slow.

To apply the JND concept to the practical evaluation of drumming time series, it is first necessary to identify what forms the background in a drumming performance. When a person strikes a drum at a prescribed tempo, essentially what they are making are inter-beat intervals. These intervals may be construed as the background against which errors are detected. The JND for drumming error at a given tempo is then found from:

$$JND(IBI) = K^* \, IBI,$$

where tempo (in beats per minute) and IBI (in seconds) are related through IBI = 60/tempo. This equation provides the practical basis for determining whether drumming errors exhibit the Weberian property of constant K. An experiment that measures JNDs at different IBIs is not difficult and it is straightforward to determine if K is a constant. And in fact, adult drummers do exhibit the constant K Weberian property at tempi greater than about 60 bpm where music is typically played (Gilden & Marusich, 2009; Gilden & Mezaraups, 2022a).

To appreciate the significance of this observation it will pay to step back just a bit and think about what Weber's law is about. Weber's law is essentially about invariance, that there is a *constant K* that relates JNDs to background magnitudes. The existence of a constant does not come about by accident; there is some organized dynamic which is producing it. There is only one way to be constant whereas there are infinite varieties of inconstancy. Constancy should always be viewed as a highly unlikely and therefore highly specialized state of affairs. When something highly unlikely is observed, that is evidence that something in the background has acted with a guiding hand. The guiding hand in nature is natural selection. The implication is that nature has given us something that allows us to operate rhythmically in the improbable regime of constant K. That something is the experience of rhythmic pulse. It is pulse that keeps us on the beat, and it is through pulse that errors are experienced. What this means practically is that Weberian behavior is diagnostic of whether pulse is being experienced.

Analysis of Drumming Time Series

We now have enough theory to penetrate how the person illustrated in Figure 8.4 experiences drumming at slow and fast tempi. The analysis begins at the bottom of the figure where IBI time series at the fastest tempi are plotted. The first observation is that the rippling amplitudes in the time series at 120 and 60 bpm are consistent with Weberian drumming, where a single constant of proportionality controls error growth. The IBI at 60 bpm (1 s) is twice that of the IBI at 120 bpm (0.5 s) and the rippling amplitude at 60 bpm appears to be about twice that at 120 bpm. At 40 bpm, however, there appears to be violations of Weber's law in that the errors appear to be larger than 3 times the size of those at 120 bpm. In addition, the 40-bpm performance does not seem be stationary in that the rippling amplitude increases about halfway through, and this increase is accompanied by a transition to a slightly slower tempo – the mean of the rippled curve moves upwards. The breakdown of Weber's law at 40 bpm is evidence that rhythmic pulse has been lost or weakened. There is no expectation that anybody could drum competently with precision at tempi below 40 bpm and it is no surprise that the time series at 30 bpm, a span of 2 s between beats, does not express pulse. This performance is not stable, starting at an IBI of 2 s and ending up at an IBI close to 3 s – a tempo of 20 bpm. This is a performance that has a wandering character; the IBIs meander as the performer slowly drifts away from where they started at 30 bpm. Wandering occurs because an IBI of 2 s is on the succession side of the phase transition for beat grouping. The drummer does not have the pulse that might guide when to hit the drum, but that the performance is wandering is informative. Wandering has a particular kind of structure, not the structure of a time series expressing pulse, but a meaningful structure, nevertheless.

Slow drift is symptomatic of a random walk, and it is interesting that lost drumming might take on this character. That a drumming performance might look like a particle diffusing in a gas or liquid suggests that, at a mathematical level, the performance is, in fact, diffusing. There are many ways of mathematically modeling diffusion, but the simplest model is sufficient to understand the onset of lost drumming. The simplest model is described by a recursive relation:

$$\text{new position} = \text{old position} + \text{random kick.}$$

What makes this recursive is that the current new position becomes the old position in forming the next new position. The psychologically relevant question is how recursion ends up describing lost drumming.

Recursion enters drumming performance through imperfect copying. A guess at what the performer is doing at 30 bpm is that, having no sense of rhythmic pulse, they attempt to recall, as best they can, their last inter-beat interval. It is important to stress here that memory of the last IBI is what might be called long-term memory. It is not the kind of activation-based dynamical memory that underlies pulse. When the activation that underlies pulse has decayed, all that is left is the explicit memory of the last IBI. The last IBI is arguably the only information available that might suggest a point in time to strike the drum. Remember that this is a person who is in an experiment where they are required to come up with a drumming performance. Given this instruction, the effort to use memory of the last IBI leads to recursion:

$$\text{new IBI} = \text{memory of last IBI} = \text{last IBI} + \text{kick,}$$

where the *kick* expresses that memory imperfection will add or subtract a little time from the true last IBI. This situation closely resembles the game of telephone where the meaning of a message drifts as it is passed, with the intention of replication, from one person to another. There is some irony here in identifying a mathematical expression for drumming in the absence of rhythmic feel. There is no mathematics for the feeling of pulse, it is an emergent property that arises when beats are perceived in relation to one another.

The time series at 15 bpm does not have a specific mathematical interpretation. There is some wandering but mostly there are also large fluctuations in the inter-beat intervals with each successive drum strike. A 4 s gap between beats turns out to be a very long time. Without the artifice of subdividing the beat by counting off seconds or fractions of seconds, this time series suggests that a person will only have a hazy notion of how long 4 s lasts. At this point not only has the feel of pulse evaporated but so has their sense of their last inter-beat interval. The only knowledge that might be useful in this regime is that they are awake, they are in an experiment, and they are supposed to strike a drum very, very, slowly.

A Collection of Pulse Phase Transitions

An experiment that measured the relationship between body size and the phase transition between rhythmic pulse and succession was conducted by

Gilden and Mezaraups (2022a). This was a simple experiment, involving nothing more sophisticated than a drum that transmitted moments of drum strike to a computer. The goal was to capture the phase transition in the formation of rhythmic pulse. Practically, this meant finding the slowest tempo that a person could drum without wandering – their watershed tempo. As the expectation was that these watershed tempi would lie somewhere between 40 and 80 bpm, it was decided to have people attempt drumming performances at the following eight tempi: 40, 45, 50, 55, 60, 65, 70, and 80 bpm. In retrospect, a wider range of tempi might have been a better choice. Drumming performances were collected from 58 undergraduates and 45 children between the ages of 6 and 12. All the performances were translated into time series of inter-beat intervals. Through trial and error, it was found that finding watershed tempi was facilitated if all the performances from a given participant were plotted together in one space on the same scale. Figure 8.5 illustrates the efforts of one participant. This person is in no way special, and these time series are offered only to give a sense of what is involved in finding pulse phase transitions.

In Figure 8.5 the labels *lost*, *pulse*, and *transition* have been added, which are justified by the following logic. The time series at 60 bpm and faster have the character of being relatively flat with rippling. The rippling amplitudes seem to be consistent with a constant Weber fraction, *K*. Further, these time series appear to be stationary in the sense that neither the rippling amplitude nor the mean level of rippling vary over the course of a given performance. The conclusion to be drawn is that this person is experiencing rhythmic pulse at 60 bpm and faster. On the other end of the tempo continuum, the time series at 40 and 45 bpm have the slowly building hills and valleys of random walks. This form of wandering is symptomatic of recursion, attempting to repeat the last IBI when the sense

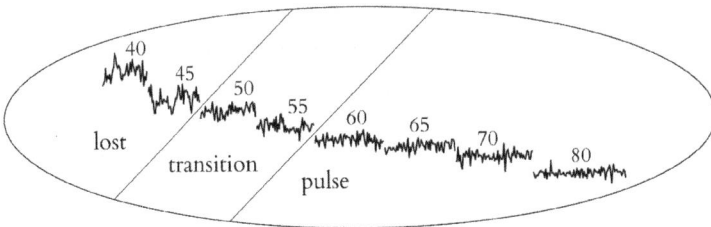

Figure 8.5 Copyright © 2022, American Psychological Association

of pulse has been lost. The two remaining performances at 50 and 55 bpm locate the region where rhythmic pulse is coming apart. These time series are intermediate between flat/rippling and wandering. It is here where the visual system's landscape contour sensitivity is most important. The performance at 55 bpm is not wandering but neither is it stationary. The second half and first half do not resemble each other, which is a symptom of pulse uncertainty. The performance at 50 bpm is difficult to characterize. It has the appearance of proto-wandering – being at the edge of wandering. It is exactly this kind of category fuzziness, not quite being one thing or the other, that is expected at a phase transition. The conclusion here is that this person has a phase transition for perceiving beats in relation to one another in the neighborhood of 1.1 to 1.2 s (the IBIs corresponding to 55 and 50 bpm).

For each of the 103 participants a plot identical to Figure 8.5 was created, and employing the exact same considerations just outlined above, each participant was assigned a watershed tempo. Importantly, the watershed tempi were assigned without knowledge of the participant's height, with the caveat that it was possible to distinguish child from adult performances on the basis of time-series length. Children drummed for 15 s whereas adults drummed for a full minute. Other than knowing whether a performance came from an adult or a child, there was no height information available and so there was no bias, conscious or unconscious, that might produce fictional allometries within either group.

The full collection of judgments of the tempi of slowest non-wandering performances together with the drummers' heights is shown in Figure 8.6. The x-axis gives the height of the participants and the y-axis gives the watershed tempo (in bpm) separating stable from wandering performances. The y-axis is also ruled with the inter-beat interval (in seconds) as it may be more useful to think in terms of time between drum strikes than in terms of strike frequency. The plot shows 103 data points, 58 filled circles showing adult participants and 45 open circles showing child participants.

Prior to a discussion of the manifest trend that exists between body size and watershed tempo, there are a few aspects of Figure 8.6 that require clarification. First, although allometries are generally expressed as power laws of mass, mass in humans is a complicated variable because it includes the hugely variable contribution of adipose tissue. A more appropriate mass variable in humans is fat-free mass.[3] Fat-free mass, however, is not easily measured. The simplest variable for the characterization of human size is height and that is what is plotted in Figure 8.6. Throughout the

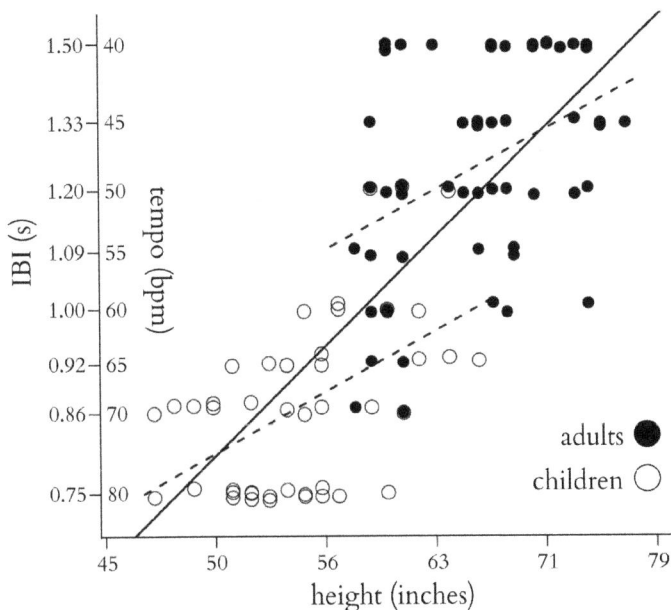

Figure 8.6 Copyright © 2022, American Psychological Association

remainder of this book allometries will be derived in terms of height simply because it is the most accessible measure of body size. The second issue is the nature of the axes. As in the analysis of mammalian action units, the data in Figure 8.6 are plotted in the log–log plane where power laws are transformed into straight lines. Again, straight lines have the virtue that they are not only the simplest structures for visual understanding, they are also the simplest structures to mathematically analyze. The third issue is that the data, the circles, appear to be laid out on horizontal tiers aligned with bpm tick marks. This is not an artifact or error, but a consequence of the judges assigning slowest non-wandering performances to actual per-formances that were recorded with no attempt to interpolate between bpm. The horizontal tiers would appear perfectly flat except that the circles were jittered a little to prevent data points from lying on top of one another, making it a little clearer where the data reside.

There is a technical problem in interpreting the child data that arises from the fact that children grow, and so height and age go hand in hand. Allometries are about body size and not about age, so this is an issue. The observation that taller/older children are better able to maintain rhythm pulse at slow tempi may simply be a consequence of maturation and overall brain development. The technique for evaluating the separate effects of age and height on the watershed tempo is known as multiple regression. Using this technique, it is relatively straightforward to show that for children in the age range sampled (6–12) the watershed tempo is determined by body size and not by age (Gilden & Mezaraups, 2022a).

Just looking at the data, the filled and unfilled circles, this plot makes a very strong case for pulse phase transition allometry. The general trend underscores a very simple point: as height increases, the tempo at which people lose their sense of rhythm decreases. Taller people can, according to these data, experience connections between beats, rhythmic pulse, at tempi where shorter people are experiencing nothing more than a succession of disconnected beats. The simplicity of this finding obscures the long road that had to be traveled before an experiment that could produce Figure 8.6 could be imagined. Figure 8.6 is both odd and unique in that it provides a space where a relation between body size and the location of a phase transition can be contemplated. There are no other figures like Figure 8.6.

Pulse Phase Transition Allometry

For almost a century body size scaling has been an area of active investigation in animal morphology, locomotion, and physiology. So, if Figure 8.6 were plotting body size versus, say, claw size or maximum acceleration or mortality, the finding that body size is a factor would not be noteworthy. Rather, scaling would be presumed, essentially as a null hypothesis. But Figure 8.6 is not drawn from biology; it is drawn from the psychology of emergent properties. That emergence and the body are related is not something that could have been presumed, and it truly is noteworthy that Fraisse's span of comprehension expresses the body. Still, there is more that might be done here. The relationship between body size and the watershed tempo comes from somewhere. It is not produced by chance so in some sense it must be lawful. The data invite us to try and figure out what that law is. The invitation is essentially to do the practical work of allometry – modeling the data in some kind of regression analysis and extracting the power law exponent that relates body size to watershed tempo. Accepting this invitation, however, inevitably involves a confrontation with the reality that even fitting straight lines to data may not be straightforward.

There is a subtle statistical problem that arises in fitting lines and extracting allometric exponents when there is more than one way of conceptualizing how the data should be partitioned into groups. This issue was discussed by Heusner (1982) in the context of the Kleiber law and his discussion is equally relevant here. Heusner's argument was essentially that the famous 3/4 Kleiber exponent was an artifact of treating mammals as a single class and that a truer picture of basal metabolism emerges when each species of mammal is allowed to have its own allometric law. Heusner's argument is illustrated in Figure 8.7.

In Figure 8.7 cartoon metabolic rates are shown for a sample of five dogs, four cats, and five mice in relation to their mass. The left panel illustrates how Kleiber would have treated the collective data. Here there is just one steeply sloped line passing through the data. The right panel illustrates the situation that Heusner was advocating where each species receives its own regression line. Now there are three regression lines and so there are potentially three different slopes and three different intercepts. Heusner argued that there was a common slope of 2/3 across species, but that each species required its own intercept. Consequently, in the right-hand panel of Figure 8.7 the regression lines have been drawn parallel to each other to reflect the common allometric exponents, and the lines have been given vertical displacements to reflect the species-specific intercepts. The insight here is that the vertical shifts translate into a steepened slope when all the data are fit with just one line. This explains how animals that on a species-by-species basis might be described by a 2/3 law could nevertheless express a 3/4 law when put into one large collection. From this point of view, there is no need to explain the 3/4 law because it is not real; it arises when intercept shifts are mixed into real but shallower allometries.

Following Heusner's treatment of the Kleiber law, there are two separate regression models plotted in Figure 8.6. One model, the solid line, is a simple regression that treats the entire data set as coming from just one

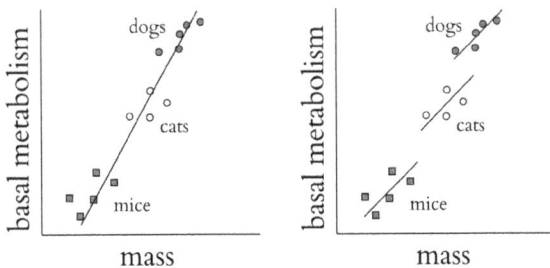

Figure 8.7

population – people regardless of being child or adult. The regression line in this model has a slope of 1.5. Remember that as the regression analysis is done in the log–log plane, the relationship between pulse and height is actually:

$$\text{pulse phase transition (s)} \sim \text{height}^{1.5}.$$

Allometries are generally not conceptualized in terms of height, presumably because most animals are not bipedal, and the heights of most animals are not as biologically relevant as their masses. To cast this allometry in terms of mass, a relationship between stature (height) and mass in human populations is needed. The accepted empirical relation dates to Quetelet (1842):

$$\text{mass} \sim \text{height}^{2},$$

where mass in this relation is most closely related to fat-free mass (FFM), there being no lawful relation between height and fat mass. The allometry for pulse may then be written:

$$\text{pulse phase transition (s)} \sim \text{FFM}^{0.75}.$$

The slope of the simple regression line in Figure 8.6 is in fact quite steep. To put the exponent of 0.75 into context, the Kleiber law predicts a 1/4 mass power law for physiological times like heart and respiration periods. The difference between 0.25 and 0.75 is not as small as it might seem because these numbers occur as powers in a power law. Because these numbers are powers the correct way to think about them is that a 0.75 law is the *cube* of a 0.25 law. The sheer size of a 0.75 exponent suggests that there may be a problem with the simple line model. Following Heusner's argument about species, perhaps children and adults should not be treated as comprising a single group.

A more refined and realistic model treats the children and adults as being members of separate groups. Although splitting the sample up into adult and child groups could have led to a child slope and a different adult slope, this did not turn out to be the case. Several independent rounds of judgment were applied to both the children and adult data in the course of developing consistent criteria for locating the slowest non-wandering performance. This process eventually led to the conclusion that the two groups shared a common slope, albeit on lines that were separated by an intercept difference. A model where children and adult pulse transitions are equally sloped, but offset, is illustrated by the dashed lines in Figure 8.6. This is exactly the situation envisaged by Heusner for metabolic allometry. A common slope means that children and adults are impacted equally by height within their respective cohorts. An intercept difference means that there is a factor besides height that influences phase transitions such that adults are overall better able to keep a sense of rhythmic pulse at slow tempi. Adults as a group seem to have a 10-bpm overall advantage over children as group. The common slope of this more refined model was 0.82 implying that:

$$\text{pulse phase transition (s)} \sim \text{height}^{0.82},$$

or recasting in terms of fat-free mass (FFM):

$$\text{pulse phase transition (s)} \sim \text{FFM}^{0.41}.$$

These relations stand as best estimates for how body size influences the divide between beat integration and beat succession. It is the first glimpse of the $\tau = f(body)$ that we have been seeking. This is the first time a relation like this has been contemplated in the history of psychology, and unlike the allometry that was developed for mammalian action units, this allometry is supported by quite a bit of data. If we take this glimpse of $f(body)$ seriously, then the exponent deserves some interpretation.

Notes

1. Musicians will subdivide intervals at very slow tempi. So, if a person is having difficulty playing a ballad in time at 50 beats per minute (bpm) they can subdivide the interval by tapping their foot at 100 bpm. Eighth notes then get one beat, quarter notes get 2 beats, and so on. This strategy works to the extent that a person's sense of rhythmic pulse is stronger at 100 bpm than it is at 50 bpm, which is generally true. In this discussion it is presumed that intervals are *not* subdivided.
2. The performances visualized in Figure 8.4 were originally recorded as part of a study of ADHD temporality (Gilden & Marusich, 2009), a precursor to the studies that motivated this book (this person did not have a diagnosis of ADHD). Each performance began with drumming along with a metronome click track at a specific tempo. After 10 clicks or so the click track turned off and the participant then continued drumming at the given tempo for another 60 drum strikes.
3. Fat mass includes adipose tissue – what is commonly referred to as fat. The amount of adipose tissue a person carries may be set by factors that go well beyond biology, including socioeconomic status, lifestyle choices, health issues, genetic predispositions, age, and the zip code where one resides.

CHAPTER 9

Allometries in Body Energetics

An allometry for a phase transition is philosophically on an entirely different footing than a biological allometry. A phase transition that separates grouping from succession is a phase transition separating two forms of experience. τ allometry inevitably refers to qualities of perception that are not reducible to the kind of physical stuff that is subject to physical law or Euclidean geometry. Although grouping, Gestalt, phase transitions, and τ can be sensibly discussed, they are truly immaterial properties of mind. It is simply not obvious how to interpret an allometry that does not refer to matter. Interpretation may, in fact, be too ambitious a goal. What is possible is contextualization, and for that purpose there may be no better point of departure than the allometries that are produced when the body consumes energy. A focus on body energetics will provide key landmarks for framing the scaling law that has been found for the pulse phase transition:

$$\text{pulse phase transition (s)} \sim \text{height}^{0.82} \sim \text{FFM}^{0.41}.$$

Body Size and Walking Energetics

The theory that leads to an allometry for walking energy usage begins with an analysis of walking efficiency. Two ways of measuring energy expenditure in walking are power, energy output per second (watts), and efficiency, energy output per meter walked. Watts are a common index of energy use in contexts related to health or strength. For example, in bicycle racing, peak or sustained watts per kilogram of body weight provides valuable information about training goals and probable race outcomes. In the commerce of ordinary life, however, the more practical measure of energy consumption is energy expended per meter traveled per kilogram of body weight – a measure of walking efficiency. Walking efficiency informs

136

on the ecologically important quantity of how much energy is used in achieving the goal of walking, moving from point A to point B. Although not an issue for people in modern industrialized cultures, energy expended in moving from place to place was enormously important for our foraging ancestors. Time acts as a vise when calories are scarce, and for most of human history walking efficiency was a routine matter of life and death.

One of the truly interesting things about walking physics is that the energy consumed in walking a given distance depends on walking speed. The relation between efficiency and walking speed is illustrated for one person in Figure 9.1.

The curve shown in Figure 9.1 is a best fit to data from a participant in a large study of real-world walking behavior (Baroudi, Barton, Cain, & Shorter, 2024). The *x*-axis gives the walking speed in meters per second and the *y*-axis gives the energy expenditure in joules per kilogram per meter walked. The general shape of the energy expense curve is parabolic, reflecting the principal fact about walking, that there is a global minimum in energy consumption. There is one walking speed that is most efficient, and for this person, that speed occurs at about 1.1 meter per second. Although people can walk faster or slower than their most efficient speed, it is a fact (Baroudi et al., 2024) that people tend to walk at a pace that is quite close to the sweet spot. The sweet spot exists because walking, when it is healthy and normal, is essentially a pendular motion, and the pendulum is a resonant system.

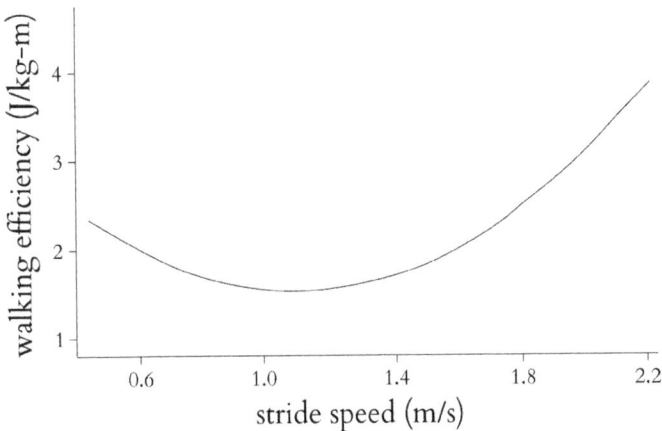

Figure 9.1

Resonance is a phenomenon found in objects such as springs and pendula that have a natural period of oscillation. The natural period of a pendulum is proportional to the square root of its length:

$$P = 2\pi\sqrt{(l/g)},$$

where P is the pendulum period, l is the pendulum length, and g is the gravitational acceleration (9.8 m/s^2).[1] To understand resonance, the formula is not as important as having a good example in mind – the playground swing, for instance. A swing with a long chain will take longer to transit to-and-fro than a swing with a short chain. Period and frequency are inversely related: the longer the swing chain the longer the period and the lower the oscillation frequency. As a playground swing inevitably transits to-and-fro at its natural frequency, it will best absorb energy if it is pushed in time with its to-and-fro period. The technical way of stating this is that resonant systems with a natural frequency are tuned to absorb energy from a periodic driver at that frequency. Now, one is always free to attempt to push a swing at whatever interval one chooses, but swings are constructed so that the person pushing is required to deliver pushes when the swing is available for being pushed, and that will occur periodically at an interval set by the square root of its length. Legs, however, are a little different than swings in that, being attached to the body, they are always available for a push, and they can be driven within a continuous range of frequencies. At high driving frequencies there is fast walking and at low driving frequencies there is slow walking.

A generic sketch of the amplitude of a damped oscillator at different driving frequencies is illustrated in Figure 9.2. Damping was included in the calculation that led to the sketch because the generic system, including the human leg, is damped by friction. The x-axis is the frequency of the periodic driving force, and the y-axis displays the motion amplitude of the damped oscillator when driven at that frequency. The most important feature of the amplitude curve is its shape, which resembles an upside-down parabola. At the resonant frequency, f_{res}, the oscillator can absorb energy most efficiently and this leads to a sharp increase in its motion amplitude. Insofar as legs are pendula and as pendula resonate at the natural frequency of oscillation, legs will have an energy absorption curve resembling Figure 9.2 that peaks at their natural frequency. Practically, what this means for walking is that when the walking speed has the legs moving back and forth at their natural frequency, the legs will be able to absorb energy most readily from muscular effort. When the legs absorb

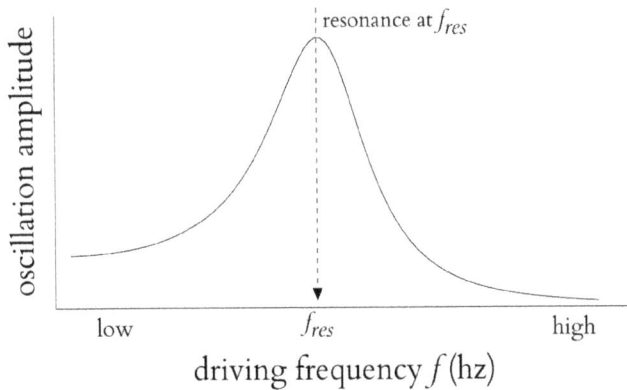

Figure 9.2

energy readily, walking is efficient. In this sense the energy consumption minimum in Figure 9.1 is the outcome of the resonance in energy transfer in Figure 9.2. To be sure, people do not have to walk at the resonant frequency, but the resonant frequency does give an energy structure to walking that is not present in any other form of locomotion. There is one speed of walking that is most efficient, and the existence and uniqueness of that one speed has psychological implications.

Resonance Theory of Phase Transitions

There are two numbers that have arisen with respect to phase transitions that need to be accounted for. One number is the characteristic value, 2±1 s, which is observed as the phase transition between group formation and group dissolution. That number set this inquiry in motion, and it has yet to be given any rationale. The second number has just arrived, and it is 0.41, the mass scaling exponent in the allometry of the phase transition in rhythmic pulse. Walking energetics provides an ecological account for the first number, the 2±1 s, through the way walking links resonance in frequency to bodily transport in space.

The argument begins by noting that the resonance in leg pendulum motion creates a characteristic crossing time, the time that it takes for a person to traverse the space that the body occupies. A crossing time, of course, depends on how fast a person walks, but when walking is maximally efficient the crossing time reflects the physics of the pendulum, and it will be about one pendular period. Now, when a mind adapts so that τ, the activation lifetime, and P, the gait period, are roughly equal, then τ becomes the crossing time for efficient walking. And when τ is the crossing time, significant activation decay occurs during the time it takes to transport the body across its own size. It is in this way that time-based grouping becomes linked to spatial displacement. When τ is tuned to the leg pendulum resonance, grouping is forced to be local to where the body is as it moves about in the environment. Events that have occurred a couple of steps in the past will have lost activation and so will not be available to group with events occurring at the current body position.

Evidence that P and τ are in fact linked is simply that the location of phase transitions is in rough numerical agreement with the time spent in a gait cycle. Taking a representative leg length of 0.8 m (an estimate of the leg pendulum length of a person who is 5' 3", the average height of women in the USA), the period in the simple pendulum approximation is:

$$P = 2\pi\sqrt{(0.8/9.8)} \approx 1.8 \text{ s.}$$

Although the leg is not a simple pendulum but a compound pendulum with a distributed mass, a more exact calculation will not change the numbers greatly. The gait period for people is inevitably going to be located around 2 s, a number seen many times before in the guise of 2±1 s, the phase transition for time-based grouping. In the most elementary terms, if *bit* is located *there*, and *ter* is located *here*, then there will be no experience of *bitter*.

There is another number to be considered here, the power of 0.41 that describes how pulse phase transitions scale with body size (cast as fat-free mass, FFM). Although a grouping phase transition is not physical, it is still of interest to inquire how its allometry aligns with the allometry that describes crossing time. For maximally efficient walking, the allometry for the crossing time comes from the physics of the pendulum:

crossing time ~ natural period of leg oscillation ~ (leg length)$^{1/2}$ ~ height$^{1/2}$.

Using the Quetelet relation (mass goes as square of height) to write this in terms of fat-free mass we obtain[2]:

$$\text{crossing time} \sim \text{FFM}^{0.25}.$$

A power of 0.25 is not the power that was experimentally derived for the pulse phase transition. That power was 0.41 when the base was fat-free mass. The issue is now whether the true power in pulse allometry is 0.25, and that it appeared in a regression analysis as 0.41 due to the large amount of unexplained variance in the regression. In other words, is the allometry so ill-determined that although the power came out to 0.41, there is so much noise in the data that 0.41 and 0.25 are not actually distinguishable? Answering this question involves estimating how much error there is in the derivation of 0.41. This is a technical issue that nevertheless makes frequent appearances in everyday life. In a political horse-race poll, for example, if one candidate has 41 percent approval and another candidate has 25 percent approval, the question that inevitably arises is whether the difference in the approval ratings is larger than the polling errors. In the public arena, the "error" is never derived or meaningfully discussed because it is technical, but the audiences for these polls still have a basic understanding of the issue.

Statistically derived quantities are always in error. If the error is due to chance and not to bias or actual mistake, it can be estimated using basic statistical principles. Error is exceptionally important in the development of allometries because a power law is only interpretable to the extent that the power is known with some precision. The error in a slope (the power in the log-log plane) is referred to in regression analysis as the standard error of the estimate. Standard errors are a measure of uncertainty that is sensitive to both sample size (number of animals) and effect size (the degree to which mass does in fact matter). In this study the standard error of the estimate is relatively small both because of the large sample size (103 people) and because height was effective at explaining watershed tempo variability. The standard error in the power estimate is 0.08, and the statistically correct way to report the power is to include it as an error bar:

$$\text{power} = 0.41 \pm 0.08.$$

The value of 0.08 has a decisive role to play in the interpretation of pulse allometry. Measured in units of 0.08, 0.41 (pulse power law) and 0.25 (pendulum power law) are far away from each other. To get from 0.41 to 0.25, chance would have to produce 2 units of movement of size 0.08 in the estimate of the power: $0.41 - 0.25 = 2 \times 0.08$. The probability of chance producing 2 standard errors' worth of movement is the sort of thing that can be looked up in a table; it is about 0.02. The implication is then that about 2 times in 100 a sample of drummers that followed a 1/4 law in their phase transitions would end up looking like they were following a 0.41 law. A value of 0.02 is considered within the culture of psychology to be improbable. The conclusion to be drawn, then, is that walking physics provides an elegant explanation of why phase transitions occur at

2 ± 1 s, but walking physics also predicts that body mass will more weakly impact the pulse phase transition than was found.

Body Size and Basal Metabolism

The second principal way that humans consume energy is through the metabolic processes that offset heat loss into the environment and so keep the body at constant temperature. Unlike walking energy usage, there is no physical analysis of the basal metabolic rate (BMR) that leads to resonance or to any other construct that might have implication for phase transitions in grouping. Nevertheless, BMR allometry provides key benchmarks by setting the scaling properties of physiological timescales such as heartbeat and respiration period. In that it is relatively straightforward to measure BMR through analysis of oxygen uptake and carbon dioxide emission, quite a bit is known about it. As a practical matter, BMR scaling in humans is not posed as a problem in allometry. Allometry is a highly restricted view of an animal property, using just the variable of mass to explain variation in that property. The human body affords many dimensions for potential analysis including fat-free mass, fat mass, age, gender, as well as a myriad of dimensions related to blood and bone content. When there are many potential factors that may influence an outcome, the appropriate statistical tool is multiple regression where each factor competes to explain variability.

Johnstone et al. (2005) conducted a multiple regression analysis of BMR using essentially the same techniques Gilden and Mezaraups (2022a) used in disentangling height and age effects in children's loss of rhythmic pulse. Recognizing the limitations of the study, 150 mostly overweight white people living in Scotland, there is still much to be learned from it. The first variable that was invited to explain BMR variability was fat-free mass (FFM). This variable was so effective, explaining 63 percent of the total variability in BMR, there was little left for any other variable to explain. Fat mass (FM) explained 7 percent of the leftovers from FFM and age explained 2 percent of the leftovers from FFM and FM. About 27 percent of the variability in BMR was unexplained variance – not explained by any of the variables that were included. The implication of this study is that BMR does in fact reduce to allometry because it is, for the most part, just a function of FFM. The slope derived from a best-fitting procedure was 0.62, giving an approximate allometric law for BMR:

$$BMR \sim FFM^{0.62}.$$

The number 0.62 has a particular meaning in that it strongly suggests that the regression is pointing to the surface area of the body through the generic geometric relation:

$$\text{surface area} \sim \text{mass}^{2/3}.$$

What is noticed here is that, numerically, the difference between 0.62 and 0.66 is not great. The Johnstone et al. study leads to the conclusion that BMR in people scales with the surface area of the body. The implication is that homeostasis is achieved when BMR offsets radiative cooling at the body's surface.

Allometric relations for physiological timescales are derived from the metabolic allometry in two steps. First BMR is recast in terms of the specific BMR, the amount of energy produced per pound of animal:

$$\text{specific BMR} = \text{BMR}/\text{FFM} \sim \text{FFM}^{0.62}/\text{FFM} = \text{FFM}^{-.38}.$$

Specific BMR, when generalized across mammals, turns out to be a variable that has great impact on an animal's life. Because the power of the specific BMR allometry is always negative, larger animals demand less energy production from each of their grams. This is a consequence of geometry. Larger animals have more of their mass sequestered in their deep interior, far away from the surface where cooling occurs, whereas the mass of smaller animals will inevitably be exposed close to the body surface. Consequently, smaller animals must organize their lives with a greater focus on acquiring calories, that is, eating.

The second step is to take the reciprocal of specific BMR to produce a quantity that has the dimensions of time. For people it may be inferred that timescales such as respiration and heartbeat periods will satisfy:

$$\text{BMR timescales} \sim \text{FFM}^{0.38} \sim \text{height}^{0.76},$$

where the Quetelet relation (mass goes as height squared) has been used to produce the height exponent. The positive exponent appearing in a timescale implies that larger people have longer heart and respiration periods. This is generally true in mammals; the issue is only in identifying the correct exponent. Humans have a surface area exponent, the same exponent that Heusner advocated for in his interpretation of the Kleiber law.

This route to timescale allometry has been indirect, coming from measurements of BMR, and not from measurements of heart rate or respiration rate. The direct route to timescale allometry is obvious; simply collect an ensemble of heart rates and heights. Although such investigations seem to be rarely reported,[3] one study that did collect a large ensemble of heart periods and heights (Smulyan et al., 1998) also computed a regression analysis that leads to:

$$\text{heartbeat period} \sim \text{height}^{0.79}.$$

The Quetelet relation allows this relation to be expressed in terms of fat-free mass:

$$\text{heartbeat period} \sim \text{FFM}^{0.40}.$$

The agreement between the direct estimate of the mass power (0.40) and the indirect estimate (0.38) is evidence that the allometry for heartbeat period can be regarded as reasonably settled.

This inquiry has landed in an interesting place. It began by noticing that gaps of 2±1 s are generally destructive to time-based grouping. That generality required a general destructive mechanism that was realized through the construct of activation decay. The construct of activation decay led to the construct of the activation lifetime, τ. As activation lifetimes do not materialize out of nothing, they must reflect properties of the system they appear in, the discussion was led toward allometry. Allometry has now led to an equality, an equality that could not have been imagined at the beginning of this inquiry: the Kleiber law that describes body-size scaling of heartbeat period also describes body-size scaling of the maximum period over which musical beats resonate in the mind. In a word, the heartbeat pulse in your neck and the rhythmic pulse in your mind share the same allometry.

We are now ready to move to the heart of temporality. There is a palpable feeling to the passage of time, it is ultimately what allows rhythm to be experienced. Although there is no real understanding of this feeling, people display the feeling of time through the organization of their behavior. Because behavior is on full display, an ethological rather than an experimental approach will allow a second, and perhaps clearer, picture of the activation lifetime and its allometry.

Notes

1. This formula is only exactly true for a pendulum bob held at the end of string of length l, not for something like a leg that has mass distributed along its entire length. A realistic mass model of a leg would not alter the square root dependence on leg length and so is not required for allometry.

2. Although this derivation of walking allometry has been based on just the physics of the pendulum, the 0.25 exponent turns out to be in remarkably good agreement with a large empirical study of animal locomotion (Cloyed et al., 2021).

3. Naively it might be thought that data on heart rate and stature would be in great supply. Although such data might be available in various databases, they do not appear to be commonly published for the purposes of correlation and regression. This study appears to be one of the few reports of a height/heart rate regression analysis in the literature.

CHAPTER 10

The Feeling of Time

This final chapter returns to the beginning, thinking about how humans and other animals experience time passage. But now we are not so naïve as to regard time passage as a prothetic continuum that makes impressions on the mind as if it were a sensory quality. The theory presented here is motivated by some phenomenology, that time can, in a limited way, be felt. There are many overt ways that people experience the feeling of time, rhythmic performance and dance being two ways that are especially compelling. There are, however, a multitude of ways that people express the feeling of time that are more quotidian and therefore less noticed. The most common of all human activities, speech, will turn out to be a fertile ground for observing how people use felt time to give their voice individual expression. To prepare for this discussion an experiment is presented that directly addresses where the feeling of time ends. An essential aspect of the time feeling is that it does, in fact, have limits. In this chapter a new kind of Stevens' law for time sense is developed that respects these limits. The new law is not a law, of course, but it will lead to new understandings of how people express their temporality.

Two Ways of Reckoning the Passage of Time

This discussion begins with an article that manages to be enormously interesting even though its conclusions may be known by absolutely everybody who is old enough to count: "When to start explicit counting in a time-intervals discrimination task: A critical point in the timing process of humans" (Grondin, Meilleur-Wells, & Lachance, 1999). This article describes a series of studies that investigated how sensitive people are to the passage of time, both when they are counting and when they are dead reckoning – just using their inner sense of time. The experiments and data analysis are technical, but the basic findings are not.

The stimuli in these experiments were simple time intervals, empty landscapes of time that began and ended with a brief burst of white noise. Intervals were presented in pairs such that one interval was a standard of fixed duration and the other was a comparison interval of variable longer duration. A trial consisted of the standard and comparison being presented in random order, and the decision to be made by the participant was whether the first interval was longer or shorter than the second. By varying the size of the comparison interval, it was possible to obtain objective measures of discrimination sensitivity. The first experiment employed a 1 s standard duration whereas the second experiment employed a 2.5 s standard.

Sensitivity in the context of discrimination is measured by a just-noticeable difference (a JND). Here the JND is practically measured by finding a difference that leads to 75 percent correct discriminations. This 75 percent comes from the circumstance that it is half-way between 50 percent (where there is no noticing and there is only guessing) and 100 percent (where there is perfection in noticing). In this way the JND is operationally defined as the number of milliseconds of difference that permits a participant to correctly identify the longer interval with 75 percent accuracy. Across the two experiments there were four conditions in what is known as a 2 × 2 design: two standard intervals (1 s and 2.5 s) and two reckoning conditions (reckoning without counting, reckoning while counting). Although it may not seem that there is much difference between 1 s and 2.5 s in the conduct of human affairs, the two standards lie on opposite sides of a fundamental divide in human time keeping. What Grondin et al. found is illustrated in Figure 10.1.

Figure 10.1 tells a compelling story, a story that effectively puts the dragons into Stevens' law. The story has two parts; what people experience when they attempt to reckon intervals of 1 s and then how that differs from the experience of reckoning intervals of 2.5 s. The first observation to be made is that when the standard comparison is 1 s, it evidently does not make any difference whether people are counting or not. Counting or not counting, the JND for deciding which of two intervals was the longer was about 100 ms, or 10 percent of the standard. Ten percent should be appreciated as evidence of good discrimination accuracy, especially in view of the circumstance that the decision is based entirely on what is remembered about the interval durations. The implication of this finding is that there is some faculty in cognition, some basic sense of time, that permits the accurate sensing of short time intervals. This sense of time is sufficiently robust that it does not require or benefit from the support that

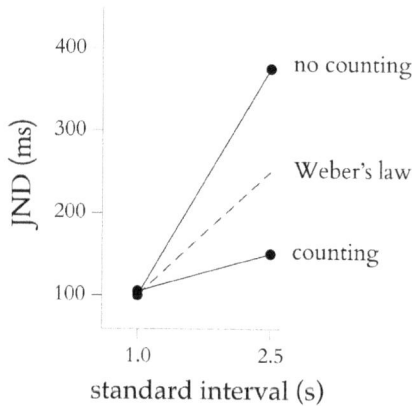

Figure 10.1

strategic counting offers. Its existence is essentially the reason that temporal duration was included within the scope of Stevens' law, and it is what pacemaker-accumulator models were designed to explain. Throughout the remainder of this discussion the thing that we have called the basic sense of time will be described more transparently in terms of what it is, a feeling of time.

The second observation to be made is that when the standard interval is 2.5 s, the JND in the no-counting condition was significantly larger than when counting was allowed. This is a situation where the numbers matter, and to put these JNDs into perspective, we should first be clear about what our expectations might be. The simplest expectation would be that duration discrimination is Weberian, that there is a constant Weber fraction (JND/standard interval size) that holds for both 1 s and 2.5 s standards. Sensory systems that are operating normally and within their design specifications will tend to be Weberian. In this case the 10 percent that was found for 1 s standards would still hold for 2.5 s standards. We now have a specific numerical prediction, that JNDs at 2.5 s would be 250 ms, and it is this number that will provide the context needed to understand the observed JNDs.

The JND in the no-counting condition was found to be 400 ms, a large enough deviation from 250 ms to warrant the implication that the feeling of time passage is not available at 2.5 s in the way that it is at 1 s. A Weber fraction that jumps from 10 percent to 16 percent (400/2,500) is a sure sign that something is broken. This is how Grondin et al. inform us that there are dragons in Stevens' law, that there are limits to how much time passage can be felt. In fact, this experiment might have been the end of treating temporal duration as a prothetic continuum that is ripe for inclusion in Stevens' law. In so far as animals other than ourselves do not count as a strategy for time keeping, it might also have led to wondering exactly what rats and pigeons are doing when they accurately respond to intervals spanning tens of seconds.

This study also quantified just how effective counting may be. The JND in the counting condition for the 2.5 s standard was only 150 ms, quite a bit smaller than the Weberian prediction of 250 ms. As a JND of 150 ms is 6 percent of 2.5 s, counting demonstrates the somewhat surprising property that it is more accurate when deployed at longer time intervals. This is definitely not how sensory systems typically work, and it raises the question of what exactly is counting and what resources it relies upon.

Counting is obviously an effective way to reckon long spans of time, even into minutes and tens of minutes. It is arguable that the reason counting works so well is that it creates rhythmic pulse, and it is rhythmic pulse that guides the count in keeping track of extended periods of time passage. Naming numbers is by itself not going to lead to accurate time keeping. The numbers must be spaced out at regular intervals, and this is something that pulse can achieve. The idea that counting is a form of rhythmic expression leads to two predictions. The first is that the growth of error in counting would mirror the growth of error in other forms of rhythmic expression. If that is true, then counting to 128 may not be much different than singing or playing a 32-bar phrase. It is the case that musicians are expected to be able to play in time and it is not unreasonable to expect that phrases of this length would conclude within a few hundred milliseconds of exact metronomic counting. Had Grondin et al. included a condition where the standard was, say, 100 s, it is likely that the JND for deciding "which interval was longer" would still be less than a second, beating the Weberian prediction of 10 s (10 percent of 100 s) by a factor of 10.

The second prediction of counting being a form of rhythmic expression is that counting would be undertaken within the same constraints that limit the formation of rhythmic pulse. Relevant to this prediction is the ethological observation that people are demonstrably not comfortable with counting slowly. It is part of growing up in the USA to discover that when people attempt to count slower than 60 bpm, they say something like *Mississippi* between each number. *Mississippi* subdivides the interval, effectively boosting the tempo into a regime where pulse is more vividly felt and where music is typically played.

In sum, these experiments teach us two things: that there is a limit to the feeling of time passage, it certainly does not extend out to 2.5 s, and that counting can achieve accurate time keeping at time intervals that exceed what might be felt. In drawing these lessons, we are obliged to refer to 2.5 s because that is the standard that was employed in the experimental design. Nevertheless, there is nothing special about 2.5 s. It is just a time interval that exceeds our capacity for feeling time and so would offer an opportunity for counting to show what it can achieve. How did Grondin et al. know that 2.5 s would be a useful standard duration? They know this in the same way that Fraisse knows that *bit* will detach from *ter* if they are separated by a couple of seconds. This is one more situation where psychologists reveal their temporality through the choices they make in designing experiments.

These two experiments in fact give us just two glimpses of a larger landscape. Sandwiched between 1 s and 2.5 s are an infinity of other possible standard intervals, and one of them is where counting first beats not counting in discrimination accuracy. Finding this one standard interval is what the title of the article is about. This interval is essentially a phase transition that is bordered on one side by a mysterious feeling process and on the other by an emptiness that can only be negotiated through counting. Grondin et al. conducted one last study to locate the phase transition. They estimate that at the standard interval of size 1.2 s, the feeling of time is attenuated to the point that subdivision through counting becomes a viable and improving strategy. It is surely no accident that this value is quite close to the metronome limit of 40 bpm where the time between successive beat divisions is 1.5 s. There are evidently connections to be made between the end of felt time and the constraints on proximity that underlay the formation of time-based groups.

The Formal Structure of Felt Time

The objective here is to create an outline of what a fully articulated theory of felt time might look like. Once we have the outline it will be possible to give the theory a precise mathematical realization. The discussion begins with a simple idea, that the feeling of time is not a static given thing, but something that changes with the passage of physical clock time. Intuitively, it seems that the feeling of time grows with the passage of clock time. This will be our point of departure, and the first issue to be addressed is what a growing feeling of time fullness might look like when expressed not as metaphor but as a graph of a mathematical function; feeling of time $= f(t)$. Like many things discussed in this book, there will be no real understanding of what felt time is, but we might be able to meaningfully discuss the dynamics of felt time and describe its law of growth.

In thinking about what the graph of $f(t)$ might look like, it will help to have a definite, concrete, situation in mind. Perhaps an appropriate vantage point might be that of a participant in Grondin et al.'s interval discrimination study. In that case we have just heard a burst of white noise and are experiencing the passage of time in the no-counting condition. The burst of noise initiates an epoch where $f(t)$ begins to grow. Conceptually, this is not substantially different than the closing of a switch in SET that starts the accumulation of counts. As the interval lengthens, the feeling of time fullness continues to grow and pass through a range of distinguishable feeling states. If we are in the 1 s condition, using this feeling we will be able to perform subtle discriminations with about a 10 percent error. But if we are in the 2.5 s condition, then we are in trouble because the feeling of time fullness does not grow without bound. Everyday experience, as well as Grondin et al.'s study, suggests that after a couple of seconds the feeling of time fullness ceases to be a reliable informant on the progression of time passage. The decoupling of the feeling of time and the progression of clock time occurs because the feeling saturates. Following the passage of a couple of seconds, time passage may still be experienced, but not as something felt. Beyond a few seconds the experience of time passage relies on episodic memory, the explicit and aware sense we have of our own autobiography. Counting is but one way in which episodic memory is used to extend the reckoning of time passage across long durations. The picture of felt time presented in Figure 10.2 shows how this works.

Figure 10.2 illustrates the two epochs that define the growth of the time fullness function, $f(t)$. Early on there is a growth phase where the physical passage of time is correlated with the growth of a feeling of time fullness. This epoch is denoted in Figure 10.2 as the regime of *felt time*, that brief period when feelings of time passage are reliable informants of physical time passage. The regime of felt time ends when the growth function turns over and saturates at an asymptotic feeling. As the feeling of time passage saturates, time passes in the world without causing a palpable change in the feeling of time. The saturating of felt time marks the boundary of a type of mental life where time passage is recognized only by the memories we have of our own lives; of the things we have done and experienced. The dragon that flew into Stevens' law of duration in Chapter 1 is now observed in its natural habitat, in the regime of episodic memory. The presence of a dragon is a signal that, beyond a couple of seconds, we must navigate the passage of time without the sure guidance offered by the feeling of time.

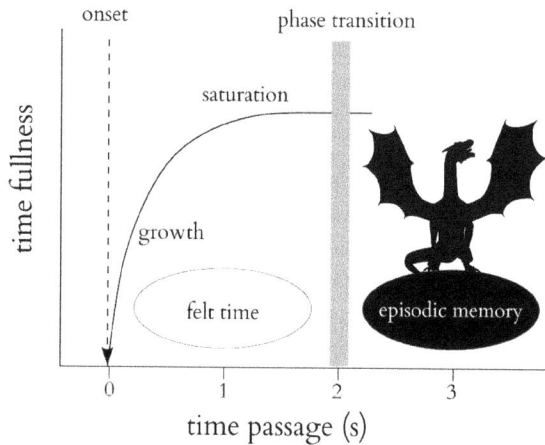

Figure 10.2

Figure/Ground Organization in Felt Time

There is one last element in Figure 10.2 to be discussed, a gray band centered on 2 s that is labeled as a phase transition. It cannot be a coincidence that the transition from feeling to counting in interval discrimination is located near the phase transition that separates time-based grouping from succession. It seems evident that there is just one phase transition, but it has two different manifestations corresponding to two different figure/ground organizations. Multistability in figure/ground organization may not be familiar and so an invitation to experience it is offered in Figure 10.3, where there are three gray disks that are surrounded by a region of black. There is more to this figure than might be imagined. First, attempt to see this illustration as a brick with three holes. If this is possible then think about how you might poke a finger into one of the holes. Now imagine touching the interior of the hole. What is your finger touching? The pad of your finger is now exploring the rough inside of the brick, and the pad is facing outwards. This is one way of experiencing the boundary between the gray disks and the black surrounding region. Now attempt to see this illustration as three stickers on a black piece of cardboard. Imagine touching the outside of the sticker. The pad of your finger now faces inwards as it explores the round curve of the sticker. After playing with this for a moment it will be possible to understand the Gestalt lesson here: whatever is figure owns the boundary. The boundary adheres either to the inside of the brick when it is the figure, or it adheres to

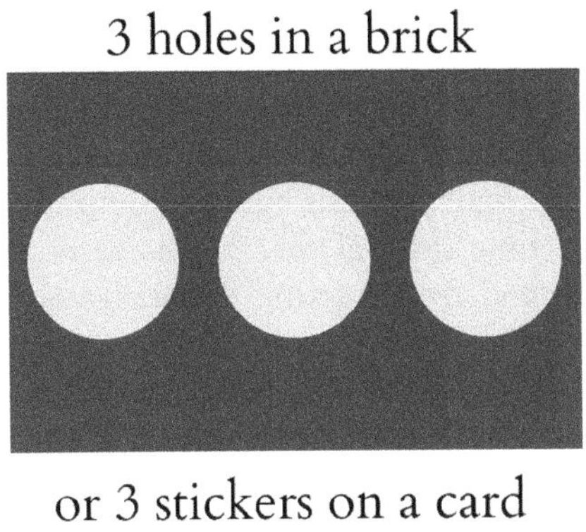

3 holes in a brick

or 3 stickers on a card

Figure 10.3

the outside of the stickers when they are the figure. The notion of adhering is quite abstract, but your body understands, and it shows this by the way the finger pads are oriented when they touch. The boundaries in Figure 10.3 evidently have dual potentiality. The phase transition in human temporality is also a type of boundary with dual potentiality.

The competing figure/ground organizations underlying our experience of time are illustrated in Figure 10.4. In an event-based description there is one event, *bit*, followed later by another event, *ter*. If the events are the figure, in the foreground of attention, then the issue arises as to how *bit* and

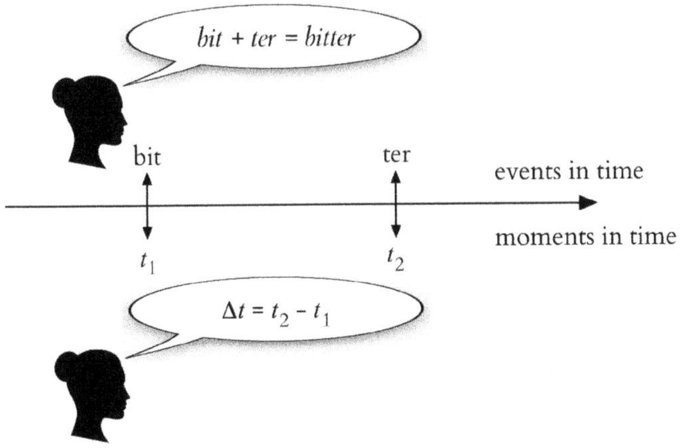

bit + *ter* = *bitter*

bit   ter   events in time

t_1 t_2 moments in time

$\Delta t = t_2 - t_1$

Figure 10.4

ter sound, whether they are heard in relation to each other, and whether they fuse into *bitter*. This is the figure/ground organization that is required to understand what Fraisse is referring to when he speculates about the capacity of apprehension. Alternatively, the interval of time bracketed by *bit* and *ter* may be the figure. In this figural organization, *bit* and *ter* act as signals that push a time interval that begins at t_1 and ends at t_2 into awareness. Now *bit* and *ter* are acting essentially as the bursts of white noise that Grondin et al. used to create time intervals. What the listener now experiences is the feeling of time passage corresponding to an elapsed time of Δt. The issue now is not how *bit* and *ter* sound, but how that bit of elapsed time feels. This is the figure/ground organization that is required of participants in an interval discrimination study.

The two different figure/ground organizations lead to two different descriptions of the phase transition in human temporality. Each description is conducted in its own specialized language. When time passage is in the background and the events that occur in time are in the foreground, the discourse is conducted in terms of a predicate that applies specifically to events, proximity. The issue in this case is whether the syllables are sufficiently proximal to be heard in relation to one another. Consequently, in this figure/ground organization the phase transition 2±1 s is conceptualized as separating a temporal regime of group formation from a temporal regime of succession. What is known about this form of the phase transition comes from everyday life and from Paul Fraisse wondering about *tick-tock* and *bitter*. When the figure/ground organization is flipped so that *bit* and *ter* are reduced to being just markers that signal the beginning and ending of a time interval, the discourse also flips and is now conducted in terms of a language specific to time intervals, a language of discrimination accuracy. So, in this reversed figure/ground organization, the same phase transition of 2±1 s is conceptualized as separating a regime where felt time permits accurate discrimination from a regime where it does not, where counting is required. What is known about this form of the phase transition also comes from everyday life and now from Grondin et al.

The idea that there is just one phase transition but two ways of describing it has implications. The allometry that was found in the phase transition for rhythmic pulse raises the possibility that there may also be allometry in some behaviors that express the feeling of time. In the following section a theory will be developed that makes this possibility explicit. At that point we will be prepared for an ethological investigation into how people show their felt time.

A Theory of Felt Time

One of the themes of this book is that psychology never proceeds from first principles. It is good to be reminded of that here because there is no fundamental theory of human temporality that will provide an equation for the growth of time fullness. Nevertheless, there are opportunities for theoretical development. In place of fundamental theory there is the picture of time fullness in Figure 10.2. A growth function that saturates is quite specific in its form and provides a meaningful target for an inquiry into what sorts of laws might explain the feeling of time passage. It turns out that this is a very easy target to hit. Saturated growth is observed when the exponential decay law describing activation decay is altered slightly to accommodate a constant inflow of fullness feeling. What follows is a brief but complete derivation of a saturated growth law that describes the immanent sense of passing time.

Starting with the decay term, recall the exponential forgetting law that was developed earlier to express activation decay:

$$\Delta A/\Delta t = -A/\tau,$$

where A denoted activation and τ is the decay lifetime. This equation describes a generic decay process where loss rate is proportional to the amount available to lose, and it leads to exponential decay. The feature of exponential decay that is most relevant here is that the decay process makes a soft landing as it approaches zero. A soft landing becomes a soft approach to the saturated experience of time fullness when a constant inflow term is added to offset decay. This identification creates a theory of felt time growth that has the somewhat odd feature that it is based on a process of decay. But something like this identification is required if time-based grouping and felt time are to share a common phase transition.

Let's call the constant rate of feeling inflow s, recognizing that what it refers to is enormously abstract and that a letter hardly describes what it is. Nevertheless, s expresses the rate at which the passage of physical time supplies the feeling of time fullness. Adding s to the decay equation and changing notation a little to make it clear that we are talking about a fullness function $f(t)$ leads to:

$$\Delta f/\Delta t = -f/\tau + s$$

which has the solution:

$$f(t) = f_\infty(1 - e^{-t/\tau}),$$

where $f(t)$ denotes the feeling of time at a moment of physical time t, and $f_\infty = s\tau$ is the saturating fullness. A graph of $f(t)$ will look exactly like the curve illustrated in Figure 10.2 where a period of growth is followed by saturation.

Conceptually the growth equation for the feeling of time fullness describes a generic dynamic where leakage is offset by filling. The example of filling a leaky bucket is an exact analogy and may be useful for understanding how leaking and filling leads to a soft approach to saturation. Leaking from a bucket has the property that the rate at which water leaves the bottom is set by the water pressure at the bottom. This pressure is set by the water height above the leak – a greater water height means more weight, which means more pressure. The water in the bucket will fill until it levels out at a height where the leaking rate matches the filling rate. This example should also make it clear that everything that is being discussed here is theoretical. All that is at stake here is the derivation of a conceptual framework that is aligned with the idea that time has a feeling and that the feeling grows until it saturates. How activation, activation decay, and inflow are represented in the brain is not remotely an issue here.

Allometry in the Experience of Time Passage

The theory of felt time has been built upon the foundation of activation decay, and so it inherits the decay lifetime, τ. This inheritance has consequences for the allometry of felt time. As τ is the portal through which the body locates the phase transition in grouping, it is now also the portal through which the body sets the phase transition where feeling saturates. Note that this is a prediction of a theory, it need not be true. There are experiments that might disconfirm it. For example, if there is allometry in felt time, then the exact experiments that Grondin et al. ran should reveal that taller people can feel time intervals that shorter people must count. That might be a fruitful direction to investigate, but a more compelling direction might be one that is truly unexpected simply because it is hidden in everyday experience.

One of the principal ways that people let their feeling of time be known is through their speech behavior, particularly in speech pauses, those moments where silence infiltrates the speech stream. Allometric influences on the experience of speech pauses will have two different presentations depending on whether the point of view is that of the listener or that of the speaker. First, consider how pauses might be experienced from the point of view of the listener. To be concrete, consider a listener who experiences a pause following the speech event:

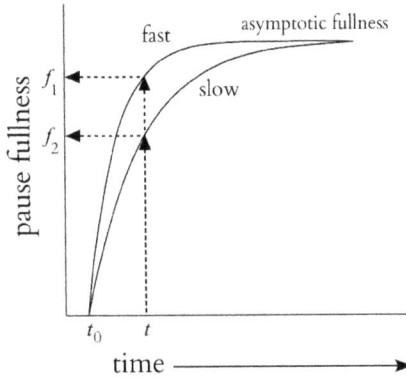

Figure 10.5

"I was, um, thinking that . . . ."

After the "that," the listener is left hanging and begins to experience the passage of time as a palpable feeling. The situation for two listeners is illustrated in Figure 10.5. One listener is tall, they might have slow activation decay, and their feeling of time fullness grows slowly. The other person is short, they might have rapid activation decay, and their feeling of time fullness grows rapidly. At $t = t_0$ the pause following "that" initiates in both listeners an epoch of feeling growth. The feeling of time passage, however, grows in these two people differently and this will have consequences for their experience of the pause.

It is important to distinguish here between time as a felt subjective experience and time as flowing in a physical sense. The speaker, by pausing, is presenting physical time passage to the listeners. In Figure 10.5 physical time is flowing along the x-axis. The subjective experience of felt time in these two listeners is read by following the arrow up from physical time and over to the y-axis where felt time is recorded as a degree of pause fullness. In this example, the divergence of the growth functions produced by body size leads to the two listeners having two different responses to the single pause that has lasted, say, for t seconds. The shorter person, having a shorter activation lifetime, will reach their point of saturation earlier than the taller person. Consequently, at time t the shorter person experiences a greater level of time fullness feeling (illustrated as f_1) than that experienced by the taller person (illustrated as f_2). As the shorter person is also closer to saturation than the taller person, the shorter person is also nearer the point at which they would have to start counting if such a pause was given as a stimulus in Grondin et al.'s discrimination study.

We now come to a subtle issue; how do these two people experience the pause? Our most direct access to time is through feel, and that is limited to relatively brief intervals. If a pause is located in the regime of felt time, then the issue is what kinds of feelings these two people are having. Insofar as the shorter person will experience greater fullness, it is sensible to conclude that they also experience the pause as being longer than does the taller person. This is a difficult proposition to test as feelings are not behaviors. They are not out in the open, available for viewing. However, when we turn the allometry around and examine how people speak, the empirical situation is decisively changed.

Consider, then, the experience of time passage from the point of view of pause creation, where people broadcast their experience of time passage. Speakers produce pauses for many reasons, and any would suffice as an example of how people use their sense of time fullness to push pauses out into the world. In this example there are two people who will be producing a pause, one short and one tall. The pause could be taken for any reason, but perhaps it is taken to mark what linguists call a grammatical boundary – a place where a comma, colon, semicolon, question mark, or period would be placed. An example of this context is the question:

What is time? Time is the . . .

It is not possible to read or speak this line in a meaningful way without pausing at the question mark. So, each speaker stops at the question mark. Then what happens?

A close analysis of what is involved in ending a speech pause makes clear that pause beginnings and pause endings have entirely different causation. Whereas speech pause beginnings might reflect linguistic structure, speech pause endings reflect temporality. In this example the question/answer context commands us to pause at the question/answer boundary. The question mark acts as a stop sign. But like a stop sign, the question mark does not tell us when to proceed with more speech. That must be decided by each speaker using their sense of time passage. This is where the time fullness function comes into operation. The anatomy of the question/answer pause is illustrated in Figure 10.6 for two speakers: one short with a fast growth rate and one tall with a slow growth rate. At $t = t_0$ each speaker stops speaking at the question mark and the feeling of time fullness begins to grow within each of them. At some point in the

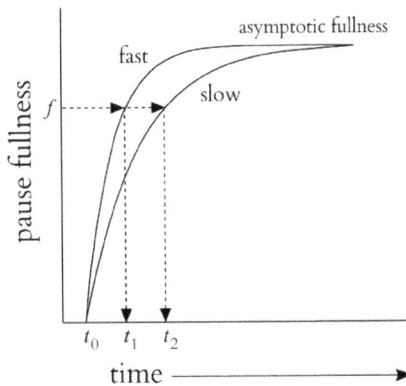

Figure 10.6

flow of physical time, the feeling of time fullness will have grown to a magnitude where each speaker feels like it is time to move on from the question mark and answer the question. When they recommence speaking, they each generate a pause whose duration can be measured. These durations are data.

To push the theory to its conclusion, we will assume that the psychological experience of time, felt time, is universal. This does not mean that everybody experiences the flow of time in the same way. It means only that everybody experiences feelings in the same way. What that means here is that if a feeling of fullness, f, is sufficient to make a short person move on from the question mark, that same feeling, f, is sufficient to also make the tall person move on. The key point is that the same feeling corresponds to two different intervals of physical time because these people have different growth functions. This is the consequence of allometry in the activation lifetime, τ. In Figure 10.6 we may then read out the consequences of allometry by following the fullness feeling, f, from the y-axis, over to the two growth functions, and then down to the x-axis where time is flowing. The shorter person has a shorter activation lifetime and consequently a steeper growth function than the taller person. Consequently, they arrive at the sense to move on from the question mark sooner. The shorter person ends their pause at t_1 whereas the taller person ends their pause a little later at t_2. In this way allometry in τ has the real-world consequence that it predicts that shorter people will exhibit faster, more compressed speech because they take shorter pauses.

An Introduction to Speech Pauses

Pauses are traditionally sorted into three classes based upon their duration. The shortest pauses, typically less than about 250 ms, are referred to as *articulatory* pauses in view of the circumstance that they are caused by constraints on articulation. An example of an articulatory pause is the brief silence, a few tens of milliseconds, that occurs between the double consonant in *happy*. This class of pauses is not of interest here because they do not involve felt time. The pauses of interest are just those that follow the stop-sign rubric where a sense of time passage must exist to recommence speech having previously stopped.

Most pauses that are ended in the regime of felt time are those where speech is briefly stopped to mark grammatical and prosodical boundaries. These pauses tend to be of intermediate length, generally between 0.25 s and about 1 s in length. In that these pauses segment the flow of speech into phrases, they perform important functions that are not strictly linguistic. As in music, phrasing in speech is how communication achieves momentum and direction. Essentially it is through these intermediate-length *segmenting* pauses that speech acquires the Gestalt property of shape and becomes personalized. Every person has a voice, and that voice is largely (and ironically) transmitted through the deployment of brief periods of silence.

There are also long pauses, longer than a second, that are taken for any number of reasons, from speech planning to distraction to checking to see if the listener is still listening. Although there is no scientific taxonomy of long pauses, they are quite interesting simply because they are long and provide an ethological perspective on what people might attempt in pause taking. One question that arises here is whether the longest pauses provide an independent measure of the regime of felt time. Would a speaker pause for so long that listeners would literally have to count if they felt compelled to keep track of the pause?

Allometry in Pause Duration

The ethology of pause durations is not difficult to conduct insofar as human speech is quite common and it is not difficult to measure pause durations. But more importantly, the analysis of pause duration allometry presents none of the subtlety encountered in the analysis of pulse phase transition allometry. It was necessary to study rhythm at the phase transition because it is only there that body size makes an impact on the experience of rhythmic pulse. It would have been much easier to study rhythm at the tempi that music is typically played at, faster than about 70 or 80 bpm, but there are no allometries where pulse is strong regardless of body size. Phase transitions are, almost by definition, difficult to study because it is at phase transitions where groups become unstable, emergent properties attenuate, and behavior loses its perceptual guidance. It is not simple to characterize a perceptual system that is being pushed to its breaking point. Speech, however, works differently. Because the feeling of time grows along a trajectory that is influenced by a decay lifetime, allometry in that decay lifetime affects every point of the trajectory. This means that every speech pause manifests allometry, regardless of how far it is from the phase transition where felt time saturates.

Extracting pauses from speech is reducible to a rote procedure and so can be automated. To appreciate what is involved, a waveform is shown in Figure 10.7. The speaker here asks: "What is time?" and then answers with "Time is the . . . ." In an acoustic waveform speech appears as thickets of white lines that denote instantaneous loudness. As pauses are moments of silence they appear as flat lines. The only substantial pause in this speech sample is that between the question mark and the answer. The other pauses have durations less than the 250 ms cut-off that is imposed to exclude the entire class of articulatory pauses. As far as the ethology of event duration goes, there is nothing simpler than cataloging speech pauses.

Two studies will be described here (Gilden & Mezaraups, 2022b; Mezaraups & Gilden, 2023). The first study was exploratory, intended only to ascertain if pause allometry was a real phenomenon. This was an open question as there were no suggestions in either the psychological or linguistics literature that pause durations would follow an allometry.

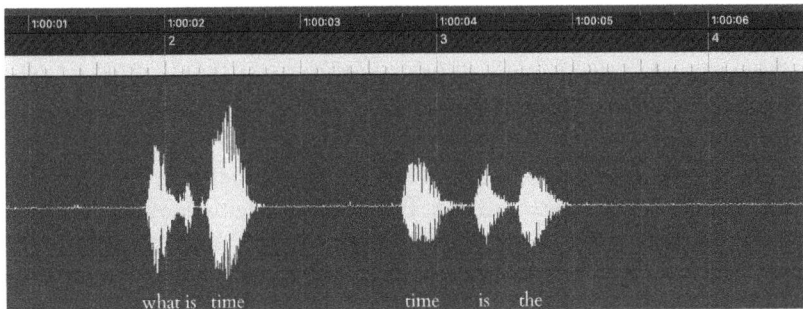

Figure 10.7

Undergraduates at the University of Texas were asked to complete a few speech tasks. These included, and again there is no scientific rationale for these choices, reading a poem, reading a paragraph, giving directions from a map, describing a cartoon, and answering questions about their major and their favorite animal. These five tasks were chosen so that pauses could be observed in several instances of both read and composed speech. The data set then consisted of the mean pause lengths in the five tasks and the heights of 68 participants.

The results from this exploratory study were remarkably clear. There was allometry in the mean pause durations for all five tasks. In the unfortunate parlance of statistics, the results were "significant" in the sense that the probability of chance producing the body-size trends were less than one in a thousand. People, it turns out, do in fact express their body size by the size of their speech pauses; taller people make longer pauses. A simple result from a simple study. Yet there is something remarkable about this discovery in view of the circumstances that led to it. Pause durations have been of interest to linguists since the 1950s and the pioneering work of Frieda Goldman-Eisler (her book, *Psycholinguistics: Experiments in Spontaneous Speech* (1968) is a classic), yet there was never a hint that pause durations might be influenced by body size. The sheer oddness of the finding of allometry raises the question: If a theoretical construct like τ implies that pause durations scale with body size, and that leads to an experiment, and that experiment supports that implication, does that mean that the theory of τ is true? No, of course it does not. This is a common logical fallacy, thinking that if A implies B, then B implies A. If B happens, it might be because of A, but it could also have been for other reasons. The theory of τ could be completely misguided and there is some sociological reason having to do with status, attractiveness, or whatever that gives taller people more pause time. Yet it is also true, nevertheless, that without this theory the experiment would not have been conducted, and the biological significance of pauses would remain unknown. There is no question that the case for this theory of τ is strengthened, but in the absence of any real understanding of what the feeling of time is, that case is hard to evaluate.

There might be more to say about pause allometry other than that taller people make longer pauses. The theory of felt time that has been developed here is based on activation decay, the same construct that was used to explain why there are phase transitions in time-based grouping. Both grouping and felt time acquire their temporal structure from a single construct, the activation lifetime, τ. That there is just one process, activation decay, and just one activation lifetime, τ, is ultimately why grouping phase transitions are located where felt time gives way to counting. The implication of a one-process theory is that body size effects that are expressed in grouping phase transitions will also be present in how time passage is experienced. The multiple roles that τ plays might be reflected in a convergence of allometric exponents.

Here we confront the difficult issue that τ is not accessible for measurement, only its manifestations. The allometric exponent that is measured in each manifestation is largely determined by the type of behavior that τ is influencing. In the context of grouping, τ makes itself known by creating a phase transition. Chapter 8 laid out the methodology for measuring the phase transition in pulse and the issue there was finding the smallest separation between beats where pulse is lost. There is just one smallest separation, and this means that the activation decay curve is being measured at just one point – the point where overlap is attenuated to the point where grouping fails to occur. In the context of speech, however, τ makes itself known with every pause taken regardless of duration. This means that the growth curve for pause fullness is being sampled at all points up to saturation. This leads to a potentially important distinction. Allometry in a phase transition is allometry in the location in real time of a single state of activation. Allometry in mean pause duration is allometry in some kind of average over activation states. It is not clear how allometry in an average quantity will compare with an allometry in a single quantity even if they have the same psychological origin in decay. At this stage of development, there is no recourse but to make the measurements and evaluate them regardless of how they turn out.

Large Studies of Natural Speech

The empirical determination of an allometric law is not a trivial matter, requiring good resolution of the behavioral property that the allometry is about. In the initial exploratory study, the undergraduate participants contributed on average about 10 pauses in each speech task. This is not a lot of data to characterize any given person's speech style. Further, each task had its own character and there were obvious differences between the pause durations in read speech and the pauses in composed speech. Read speech tends to be much more compressed than composed speech, with shorter pauses overall. The upshot was that there was little to conclude from this study beyond the finding that pause durations do scale with body size. There was no evidence for any one allometric law and certainly there was no evidence that pause durations scale with body size in the same way as the pulse phase transition. This is the nature of an exploratory study.

A larger experiment was indicated, one large enough to permit a meaningful measurement of the allometric law for the experience of time passage. Such an experiment would necessarily involve a lot more speech. Although there is no theory that mandates how a speech study should be conducted, it seemed that ordinary communicative speech might be preferable to speech issuing from a contrived prompt (read this poem, describe this cartoon . . .). It is fortunate that we live in an epoch where enormous reservoirs of recorded communicative speech are available for study. There are, at this writing, many millions of hours of video on YouTube with about three quarter million hours being added daily. Much of that content consists of people talking in natural and unscripted ways and would be appropriate for speech analysis. There is, however, the requirement that the body size of the speakers must be known if allometry is to be contemplated, and that severely limits the content that might be useful. There are two cohorts of people whose heights are published. Athletes' heights are generally well known in view of the importance that height plays in athletic achievement. Celebrities' heights are also well known and published, presumably an outcome of nosiness and fan worship. These two cohorts were examined in separate studies with pause extraction conducted by different sets of research assistants.

Mezaraups and Gilden (2023) analyzed a large corpus of interviews from some 60 athletes. Although the interviewees tended to be young, they were much more heterogeneous in every other demographic dimension than the undergraduate subject pool at the University of Texas. The interviews were generally long enough that over a few minutes of speech the athletes produced well over 50 pauses that could be analyzed. The entire data set consisted of over 3,000 pauses. Following the publication of this work, the author repeated the athlete study except now with 51 celebrities (people with name recognition in popular culture). This group was heterogeneous in every respect, age, ethnicity, gender, and, of course, height. This data set consisted of over 2,100 pauses with each celebrity contributing an average of 35 pauses.

Both studies showed clear evidence of body-size scaling, and so each produced an allometric height exponent. The athlete exponent was 0.81 ± 0.26 and the celebrity exponent was 0.73 ± 0.11. Figure 10.8 illustrates the regression of the mean pause duration for both athletes and

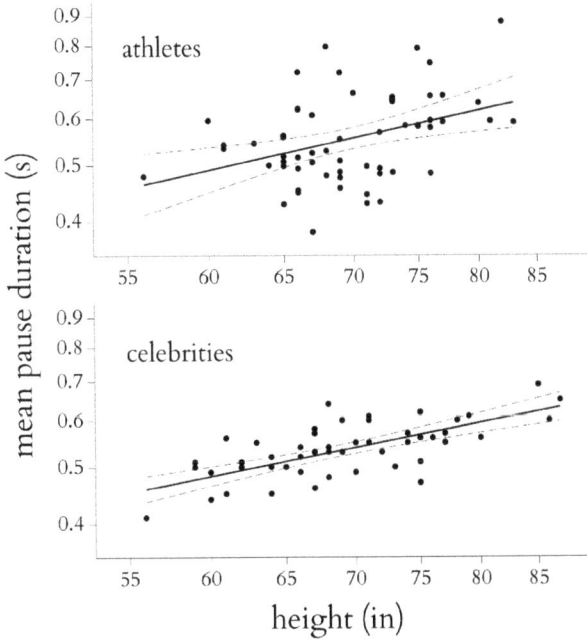

Figure 10.8

celebrities. As elsewhere where allometry has been illustrated, the axes are logarithmic so that the power law of body size becomes a linear relationship. The solid lines are regression fits, and the dotted lines show the 95 percent confidence limits (there is a 95 percent chance that the true regression lies between these bounds). In both studies the probability that the positive allometry arose by chance was less than one in a thousand.

Final Thoughts

In this book we have been collecting allometric exponents for temporal phenomena. Three exponents were empirically measured: one exponent for the pulse phase transition and one exponent each for athlete and celebrity pause durations. These exponents have been interpreted in terms of activation decay, but regardless of the theoretical background, they stand on their own as empirical outcomes. In addition, two exponents were derived from the principal ways people consume energy, through walking and through basal metabolism. An analysis of

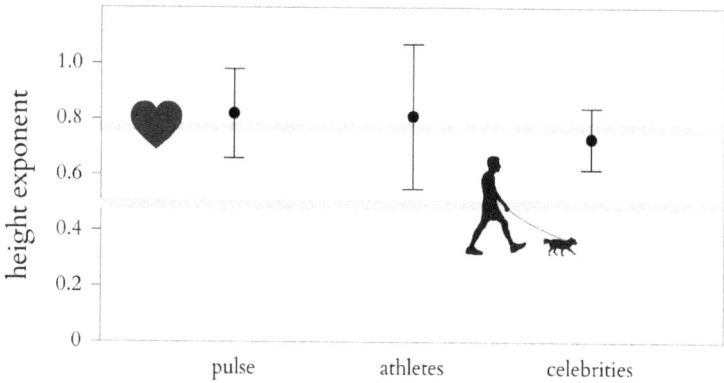

Figure 10.9

walking physics led to an exponent for the body crossing time whereas observational data from two separate studies converged on an exponent for heartbeat period. The collection is illustrated in Figure 10.9 where walking and heartbeat exponents are visualized as lines while the empirically derived exponents are visualized as data points with error bars. The case that is made in Figure 10.9 is intended to be self-evident. The three empirically derived exponents all intersect the line that is set at the heart period exponent. The walking exponent is illustrated by a line that does not intersect even the error bars. Figure 10.9 summarizes all the empirical work presented in this book.

Figure 10.9 presents an interesting picture. The empirically derived exponents are not exactly determined, but still, all three do sit on a line defined by the heart period exponent. It seems as if we are being persecuted by yet another number, this time by a power law exponent. The specific value of the heart period exponent is critical to understanding what any of this might mean. This exponent comes from an analysis of basal metabolism (Johnstone et al., 2005) that was consistent with a 2/3 law for the scaling of metabolic rate with fat-free mass. A 2/3 law has a particular meaning because 2/3 is the scaling of surface area with mass. The physiological implication is that human heartbeat period is set by radiative cooling from the body surface. There must also be implications for the exponents that sit on the heartbeat line.

This book has been about essentially just one thing: activation decay. Without activation decay there are no phase transitions and without activation decay there is no felt time. Activation decay is the central construct in human temporality. Decay creates a kind of crisis where activation is given a finite lifetime, a moment. This moment lasts only a few seconds but those few seconds are a fixture in the life of mind, and we are quite familiar with it. It is our *now*. This moment, what has been referred to more formally as the activation lifetime, τ, is part of our psychology. It also seems to be part of our body. When we measure the aspects of mind that τ controls, grouping and the feeling of time, we find allometries. These allometries implicate the surface of the body. It is as if the mind returns to a state of rest by radiating activation off the body and into the world. With this puzzle we come to a good place to end.

References

Baddeley, A. D., & Hitch, G. (1974). Working memory. *Psychology of Learning and Motivation, 8*, 47–89. http://dx.doi.org/10.1016/S0079-7421(08)60452-1.

Baddeley, A. D., Thomson, N., & Buchanan, M. (1975). Word length and the structure of short-term memory. *Journal of Verbal Learning and Verbal Behavior, 14*, 575–589. https://doi.org/10.1016/S0022-5371(75)80045-4.

Baroudi, L., Barton, K., Cain, S., & Shorter, K. (2024). Understanding the influence of context on real-world walking energetics. *The Journal of Experimental Biology, 227*(13), jeb246181. https://doi.org/10.1242/jeb.246181.

Beeli, G., Esslen, M., & Jäncke, L. (2005). When coloured sounds taste sweet. *Nature, 434*, 38. https://doi.org/10.1038/434038a.

Bregman, A. S. (1990). *Auditory scene analysis: The perceptual organization of sound*. The MIT Press.

Cloyed, C. S., Grady, J. M., Savage, V. M., Uyeda, J. C., & Dell, A. I. (2021). The allometry of locomotion. *Ecology, 102*(7), Article e03369. https://doi.org/10.1002/ecy.3369.

Ebbinghaus, H. (1948). Concerning memory, 1885. In W. Dennis (Ed.), *Readings in the history of psychology* (pp. 304–313). Appleton-Century Crofts. https://doi.org/10.1037/11304-034.

Feldhütter, I., Schleidt, M., & Eibl-Eibesfeldt, I. (1990). Moving in the beat of seconds: Analysis of the time structure of human action. *Ethology & Sociobiology, 11*(6), 511–520. https://doi.org/10.1016/0162-3095(90)90024-Z.

Fraisse, P. (1978). Time and rhythm perception. In E. Carterette & M. Friedman (Eds.), *Handbook of perception VIII* (pp. 203–254). Academic Press.

Gerstner, G. E., & Goldberg, L. J. (1994). Evidence of a time constant associated with movement patterns in six mammalian species. *Ethology and Sociobiology, 15*(4): 181–205. http://hdl.handle.net/2027.42/31468.

Gibbon, J. (1977). Scalar expectancy theory and Weber's law in animal timing. *Psychological Review, 84*(3), 279–325. https://doi.org/10.1037/0033-295X.84.3.279.

Gibbon, J., Church, R. M., & Meck, W. H. (1984). Scalar timing in memory. *Annals of the New York Academy of Sciences, 423*, 52–77. https://doi.org/10.1111/j.1749-6632.1984.tb23417.x.

Gilden, D. L. (2001). Cognitive emissions of 1/*f* noise. *Psychological Review, 108*, 33–56. https://doi.org/10.1037/0033-295x.108.1.33.

Gilden, D. L., & Marusich, L. R. (2009). Contraction of time in attention-deficit hyperactivity disorder. *Neuropsychology, 23*(2), 265–269. https://doi.org/10.1037/a0014553.

Gilden, D. L., & Mezaraups, T. M. (2022a). Allometric scaling laws for temporal proximity in perceptual organization. *Psychological Review, 129*(3), 457–483. https://doi.org/10.1037/rev0000307.

Gilden, D. L., & Mezaraups, T. M. (2022b). Laws for pauses. *Journal of Experimental Psychology: Learning, Memory, and Cognition, 48*(1), 142–158. https://doi.org/10.1037/xlm0001103.

Goldman-Eisler, F. (1968). *Psycholinguistics: Experiments in spontaneous speech.* Academic Press.

Grondin, S., Meilleur-Wells, G., & Lachance, R. (1999). When to start explicit counting in a time-intervals discrimination task: A critical point in the timing process of humans. *Journal of Experimental Psychology: Human Perception and Performance, 25*(4), 993–1004. https://doi.org/10.1037/0096-1523.25.4.993.

Hasher, L., Zacks, R. T., Stoltzfus, E. R., Kane, M. J., & Connelly, S. L. (1996). On the time course of negative priming: Another look. *Psychonomic Bulletin & Review, 3*(2), 231–237. https://doi.org/10.3758/BF03212424.

Heusner, A. A. (1982). Energy metabolism and body size. I. Is the 0.75 mass exponent of Kleiber's equation a statistical artifact? *Respiration Physiology, 48*(1), 1–12. https://doi.org/10.1016/0034-5687(82)90046-9.

Johnstone, A. M., Murison, S. D., Duncan, J. S., Rance, K. A., & Speakman, J. R. (2005). Factors influencing variation in basal metabolic rate include fat-free mass, fat mass, age, and circulating thyroxine but not sex, circulating leptin, or triiodothyronine. *American Journal of Clinical Nutrition, 82*(5), 941–948. https://doi.org/10.1093/ajcn/82.5.941.

Kleiber, M. (1932). Body size and metabolism. *Hilgardia, 6*, 315–353.

Koffka, K. (1935). *Principles of Gestalt psychology.* Harcourt, Brace & World.

Lettvin, J. Y., Maturana, H. R., Mcculloch, W., & Pitts, W. (1959). What the frog's eye tells the frog's brain. *Proceedings of the IRE, 47*, 1940–1951.

Lovatt, P., Avons, S. E., & Masterson, J. (2000). The word-length effect and disyllabic words. *The Quarterly Journal of Experimental Psychology A: Human Experimental Psychology, 53*A(1), 1–22. https://doi.org/10.1080/713755877.

Meyer, D. E., & Schvaneveldt, R. W. (1971). Facilitation in recognizing pairs of words: Evidence of a dependence between retrieval operations. *Journal of Experimental Psychology, 90*(2), 227–234. https://doi.org/10.1037/h0031564.

Meyer, D. E., Schvaneveldt, R. W., & Ruddy, G. W. (1972). Activation of lexical memory. Paper presented at the meeting of the Psychonomic Society, St. Louis, Missouri, November 1972.

Mezaraups, T. M., & Gilden, D. L. (2023). More laws for pauses: Replication and generalization. *Journal of Experimental Psychology: Learning, Memory, and Cognition, 50*(1), 161–168. https://doi.org/10.1037/xlm0001280.

Miller, G. A. (1956). The magical number seven, plus or minus two: Some limits on our capacity for processing information. *Psychological Review*, *63*(2), 81–97. https://doi.org/10.1037/h0043158.

Neill, W. T., & Westberry, R. L. (1987). Selective attention and the suppression of cognitive noise. *Journal of Experimental Psychology: Learning, Memory, and Cognition*, *13*(2), 327–334. https://doi.org/10.1037/0278-7393.13.2.327.

Peterson, L., & Peterson, M. J. (1959). Short-term retention of individual verbal items. *Journal of Experimental Psychology*, *58*(3), 193–198. https://doi.org/10.1037/h0049234.

Roediger, H. L., & McDermott, K. B. (1995). Creating false memories: Remembering words not presented in lists. *Journal of Experimental Psychology: Learning, Memory, and Cognition*, *21*(4), 803–814. https://doi.org/10.1037/0278-7393.21.4.803.

Schleidt, M., Eibl-Eibesfeldt, I., & Pöppel, E. (1987). A universal constant in temporal segmentation of human short-term behavior. *Naturwissenschaften*, *74*(6), 289–290.

Smulyan, H., Marchais, S. J., Pannier, B., Guerin, A. P., Safar, M. E., & London, G. M. (1998). Influence of body height on pulsatile arterial hemodynamic data. *Journal of the American College of Cardiology*, *31*(5), 1103–1109. https://doi.org/10.1016/s0735-1097(98)00056-4.

Stevens, S. S. (1957). On the psychophysical law. *Psychological Review*, *64*(3), 153–181. https://doi.org/10.1037/h0046162.

Voss, R. F., & Clarke, J. (1978). "1/f noise" in music: Music from 1/f noise. *Journal of the Acoustical Society of America*, *63*, 258–263. https://doi.org/10.1121/1.381721.

Warren, R. M., Gardner, D. A., Brubaker, B. S., & Bashford, J. A. (1991). Melodic and nonmelodic sequences of tones: Effects of duration on perception. *Music Perception*, *8*(3), 277–289. https://doi.org/10.2307/40285503.

Wittmann, M. (2016). *Felt time: The psychology of how we perceive time* (E. Butler, Trans.). MIT Press.

Index

For EU product safety concerns, contact us at Calle de José Abascal, 56–1°,
28003 Madrid, Spain or eugpsr@cambridge.org.

www.ingramcontent.com/pod-product-compliance
Ingram Content Group UK Ltd.
Pitfield, Milton Keynes, MK11 3LW, UK
UKHW021920211125
465270UK00008B/162